Contemporary Trends and Issues in Science Education

Volume 60

The book series Contemporary Trends and Issues in Science Education provides a forum for innovative trends and issues impacting science education. Scholarship that focuses on advancing new visions, understanding, and is at the forefront of the field is found in this series. Authoritative works based on empirical research and/or conceptual theory from disciplines including historical, philosophical, psychological and sociological traditions are represented here. Our goal is to advance the field of science education by testing and pushing the prevailing sociocultural norms about teaching, learning, research and policy. Book proposals for this series may be submitted to the Publishing Editor: Claudia Acuna E-mail: Claudia.Acuna@springer.com

Bahadir Namdar • Engin Karahan
Editors

Socioscientific Issues Focused Teacher Education

Place-Based Practices from Türkiye

Springer

Editors
Bahadir Namdar
Department of Mathematics and Science Education, Faculty of Education
Ege University
Izmir, Türkiye

Engin Karahan
Department of Mathematics and Science Education, Faculty of Education
Middle East Technical University
Ankara, Türkiye

ISSN 1878-0482 ISSN 1878-0784 (electronic)
Contemporary Trends and Issues in Science Education
ISBN 978-3-031-55232-8 ISBN 978-3-031-55233-5 (eBook)
https://doi.org/10.1007/978-3-031-55233-5

This Springer imprint is published by the registered company Springer Nature Switzerland AG
The registered company address is: Gewerbestrasse 11, 6330 Cham, Switzerland

Paper in this product is recyclable.

To
Pelin Ece and Aras Deniz

Preface

As young and passionate Turkish doctoral students at the University of Georgia and the University of Minnesota, our conversation about meaningful socioscientific issues (SSI)-based teaching and learning began in 2012 at the NARST International Conference in Indianapolis, IN. Namdar's doctoral dissertation research focused on the use of technology in socioscientific argumentation processes, whereas Karahan investigated how teachers design socioscientific issue-based lesson plans as well as their students' socioscientific reasoning. After completing our dissertations, we returned to Türkiye with a renewed enthusiasm for SSI-based instruction.

Earlier in our careers, we oversaw several professional development programs for both in-service and preservice teachers. During one of our late-night conversations after a long day of professional development, we realized that the teachers were unmotivated when we used examples of melting glaciers, but they were ecstatic to debate the construction of hydroelectric power plants in Northeastern Türkiye, a highly contentious local issue. We found it intriguing at the time that, despite climate change being a significant and current topic, the instructors' motivation and involvement levels were influenced by how close the issue was to their personal lives. Later, we launched a professional development project funded by Scientific and Technological Research Council of Türkiye to improve science teachers' argumentation skills on socioscientific issues specific to northeastern Türkiye. However, we had difficulty locating adequate resources to assist science teachers in their work.

Since then, more research has been conducted on local socioscientific issues, and scholars have formally advocated for students to have access to place-based learning opportunities. In addition to laying out current practices in the field, we hope that this book will serve as a resource for teacher educators and teachers in Türkiye and around the world who want to integrate place-based learning effectively.

We would like to thank all of the contributing authors and the preservice teachers who participated in the studies.

İzmir, Türkiye Bahadır Namdar

Ankara, Türkiye Engin Karahan

Introduction

Incorporating socioscientific elements into science lessons would help students achieve academic and nonacademic goals, contributing to functional scientific literacy (Herman et al., 2021). However, contextual elements affect the successful as well as meaningful implementation of these issues in the science learning and teaching processes, which in turn would lead to desired learning outcomes. It is important to note that not all SSI-based instructional contexts provide the targeted learning outcomes. In the context of SSI instruction, Herman et al. (2021), for example, distinguish "impoverished" and "rich" instructional contexts. Based on the relevant literature, they note that impoverished SSI contexts fail to provide students with meaningful access to the communities and environment affected by the SSI, which hinders students from making meaningful connections with the issue and forming a solid understanding of a given SSI. On the other hand, rich SSI instructional contexts allow learners to be enculturated within the community, understand the perspectives and feelings of the people affected by the issue, and feel an emotional attachment to the environment and the community. Herman et al. (2021) further argue that

> Important to note, learners will not spontaneously engage in real-world SSI in this manner. Rather, it requires that teachers implement sound research-based pedagogical decisions such as setting up concrete experiences in nature and with community members dealing with SSI, and then reflectively deconstructing those experiences through modeling thought processes and modes of inquiry, asking highly effective questions, strategic grouping, discourse, and role-play activities. The results of creating rich contextual learning environments are crucial for promoting scientific literacy for the general public by providing opportunities for learners to interpret science within the framework of real-world problems, critically analyze information, and integrate knowledge gained from both formal and informal experiences. (p. 590)

Therefore, providing rich contextual SSI environments becomes a pressing issue as Zeidler et al. (2019) also argued solely engaging students in formal learning contexts fell short of attaching students to the lived experiences and making students not see the importance of solving issues for the sake of local and global environment; engaging in meaningful practices to understand SSI; and demonstrating character and values such as ethical and moral sensitivity to resolve SSI.

Evidently, SSIs have been studied less locally than globally, and many global SSIs originate from local concerns that grow globally (Capkinoglu & Yilmaz, 2018). In other words, every global issue is linked to a local issue. In the realm of education for twenty-first-century citizenship, learners are expected to feel as part of a global community and be prepared to take action and make well-informed decisions on these ill-structured issues that span from personal to local and global scales (Lee et al., 2013). Hence, bringing local issues as "rich" SSI instructional contexts engages students in personally relevant learning experiences and provides a space for considering the global effects of decision-making, transferring local knowledge to the global. Therefore, bringing local and global elements into learning contexts becomes evident in the global citizenry, and glocalized learning environments are suggested to provide learning spaces for thinking globally and taking actions to solve SSI locally (Caniglia et al., 2018; Tien & Talley, 2012).

Glocalization approaches emphasize the importance of "place" (Caniglia et al., 2018), and place-based experiences are essential for students to take action to solve SSI facing their community and learn about scientific knowledge and practices associated with the decision-making processes (Birmingham & Calabrese Barton, 2014). Furthermore, students would be more motivated to work on socioscientific issues (SSI) that are relevant to them and close to their homes (Karahan & Roehrig, 2016) because the place offers a powerful context for learning, fostering a sense of belonging and connection as well as stewardship toward the environment (Gruenewald, 2003). Calabrese Barton and Tan (2010) conclude that the development of scientifically literate individuals is related to the "place." As a result, it is critical to provide students with learning experiences centered on local SSI to develop scientifically literate citizens.

From an SSI perspective, the place-basedness and relevancy of SSI learning environments are crucial components necessary for scaffolding students to more sophisticated ways of conceptualizing and resolving SSI (Zeidler et al., 2019). Place-based practices engage students and communities in discussions about locally relevant issues and educate community members on the actions to be taken to preserve natural resources and the environment (Birmingham & Calabrese Barton, 2014). As Karahan (2022) argued, when the students are provided with localized and place-based issues that they are familiar with, they will feel agency toward the solution of the problem. Place-based education, on the other hand, allows us to study local places in their cultural ecosystem while providing a learning space for students (Mathews, 2013).

There has been increasing interest in place-based pedagogies in SSI-based teaching and learning. Zeidler et al. (2019) synthesized recent trends in SSI research that are advanced by three broad and interrelated areas of research in their seminal work. SSI and the Importance of Place-Based Contexts was one of those research trends. Recent studies focus on the impact of place-based SSI instruction mostly on school students and including their environmental NOS views (Herman et al., 2023) and other affective domains such as compassion, intent, and action (Herman, 2018), taking educational action with the community involvement (Birmingham & Calabrese Barton, 2014), evidence-based reasoning (Powell et al., 2021), and sense of place and character development (Kim et al., 2020). However, some argue that rich place-based SSI instructional contexts are still lacking in teacher education programs. This book fills a gap in the literature by focusing on the importance of firsthand experience with this pedagogy among future science teachers.

About This Book

Although place-based education improves students' explanations, nature of science perspectives, and engagement (Herman et al., 2013, 2019, 2023), our own observations of Turkish science teachers' in-class implementation of SSI, as well as the current literature of international science education research, revealed that although teachers acknowledge its benefits, they highlight various challenges of place-based education, including meeting curricular demands as well as logistical hurdles associated with the instructional contexts (McVittie et al., 2020; Miller & Twum, 2017). Despite certain challenges, we believe the next generation of teachers will be the active agents of change in the classroom and their communities. We agree with Topçu et al. (2011) that effective SSI implementation necessitates the involvement of the next generation of teachers. We hope this book provides empirical evidence on the effectiveness of place-based SSI practices incorporated at the university-level teacher education practices and offers practical suggestions and examples from place-based SSI interventions to science teacher educators.

Second, most research on place-based SSI has focused on issues in the United States (Cook & Buck, 2014; Herman, 2018; Herman et al., 2020, 2021, 2023; Powell, 2021). As a country straddling Asia and Europe, Türkiye provides a unique context for studying place-based SSI. For example, the country's rapid infrastructure development, combined with rising energy demand, sparks public debates on a wide range of SSI, from organic agriculture to hydroelectric power plant construction, farmland use in earthquake zones for construction purposes, and sea-filled airports. These SSI attract public attention and spark ongoing debates about various issues. According to multiple review studies, energy-related SSI are widely studied in Türkiye (Evren Yapicioglu & Atabey, 2020). According to a recent report, Türkiye's rapid economic and population growth prompted the country to meet rising energy demand by shifting to renewables (International Energy Agency, 2021). Renewable energy resources are provided nationwide, including hydroelectric power in the northeast, solar power in the west and south, and wind power in the north and west. As a result, each region presents unique SSI that capture the public's attention and spark debates.

This book project outlines effective practices for teaching SSI in Türkiye through place-based pedagogies. Each chapter describes teacher education practices and outcomes related to an SSI topic encountered in the author's region (for example, gold mining, nuclear power plant construction, and river pollution). As a result, the content is structured to allow for a thorough presentation of these practices and outcomes.

The subsequent chapters in this book add to the body of literature by introducing place-based pedagogies within the framework of SSI in Türkiye. In summary, the first chapter offers the theoretical underpinnings necessary to comprehend place-based SSI. The SSI-based teaching and learning process in a range of place-based learning environments is examined in the chapters that follow. Sense of place, socioscientific orientations, argumentation, environmental awareness, character and values, conceptual understanding, perspective taking, nature of science, emotive reasoning, and co-designing competencies are among the few of the cross-cutting problems that are covered in the book. Chapter 12 returns to the preservice teacher education for place-based SSI following the chapters that offer empirical evidence on place-based SSI training. As we wrap up the book, we offer potential venues for additional SSI practice and research.

In Chap. 1, Engin delves into the theoretical underpinnings of applying place-based approaches to SSI. First, the chapter contends that traditional science learning experiences do not effectively engage students in the sociocultural, moral, and affective domains of SSI. The chapter advocates for engaging students in place-based learning experiences to address this challenge and promote scientific literacy. According to the chapter, place-based pedagogies strengthen students' connections to their community by increasing a sense of place and ownership in resolving specific SSI. The story concludes that place-based pedagogies in SSI-based teaching and learning provide rich contexts for students' meaningful engagement with their communities and that teacher educators should use these issues to prepare aspiring educators to transform their communities and use them as "learning labs."

Chapter 2 focuses on how a semester-long course affects preservice teachers' sense of place. Nilay and Büşra investigate the hydroelectric power plant issue in Giresun, northern Türkiye, using practical and theoretical classes with their preservice teachers. Notably, it would be interesting to see an increase in their sense of place identity and the various dimensions incorporated into their responses to open-ended questions about their sense of place.

In Chap. 3, Süleyman and Selçuk focus on the issue of air pollution in Ankara, Türkiye, and argue that long-term solutions to these contentious issues necessitate a shift from anthropocentric to more environmentally conscious mindsets. This journey shows how preservice teachers' socioscientific orientations change due to place-based learning activities using naive and informed perspectives provided by the researchers. It is also worth noting a contrast between preservice teachers' choices and their actual explanations of the contentious issue presented.

Another local energy issue is the development of a nuclear power plant in Sinop, Türkiye. In Chap. 4, Nurhan and Ayşe argue that making conscious decisions about this complex SSI helps to raise awareness and improve learners' argumentation skills. They address the issue by engaging preservice teachers in argumentation through an expert-supported teaching program, in which preservice teachers are given opportunities to explore the issue's multifaceted nature by discussing it with experts who have different ties to and positions on it. The chapter concludes that participation in place-based activities improved preservice teachers' arguments and knowledge transfer.

In Chap. 5, Ahmet and Mehmet focus on the pollution of the Porsuk River in Eskişehir, Türkiye, with a particular emphasis on raising awareness among preservice primary school teachers. The chapter delves into the details of pollution and highlights the proposed design-based solutions via structured discussions and argumentation. The authors conclude the chapter by advocating for increased awareness of environmental SSI in order to promote active engagement and a better environment.

Ayşegül, Bahadır, and Fulden combine two research areas in Chap. 6: creative drama and character and values. According to this group of researchers, drama activities allow learners to form affective connections, which can help develop character and values for informed decision-making. The chapter addresses an important issue: agricultural land use for settlement in earthquake zones in Izmir, Türkiye. The chapter focuses on engaging students in role-playing activities and investigating how they embody character and values.

Chapter 7 explores population growth and its effects on various areas in Istanbul, Türkiye. Seda, Çiğdem, and Filiz examine how place-based learning impacts preservice teachers' conceptual understanding of the issue. In this study, preservice teachers investigated unplanned urbanization, air pollution, water resources, and noise pollution in light of population growth concerns. Following place-based activities in each subtopic, preservice teachers were instructed to engage in design-based activities to solve the problem. The chapter provides opportunities to investigate the intersection of problem-based and place-based learning approaches.

Chapter 8 examines place-based instruction for increasing energy demand in Muğla, Türkiye. In this chapter, Nilay and Hakan explain how they engage preservice teachers in perspective-taking activities. They argue that perspective-taking is critical for developing critical thinking skills, investigating the benefits of understanding different points of view, and empowering students to engage in meaningful decision-making and position-taking. The chapter also emphasizes the value of role-playing activities for perspective-taking.

In Chap. 9, Banu and Tuğba explore the relationship between preservice teachers' understanding of science and place-based SSI instruction. They argue that immersive experiences provided by place-based pedagogies can help them refine their NOS perspectives. The intervention in this chapter focuses on the sea-filled airport issue, and the findings show that place-based learning improved preservice teachers' understanding of science.

In Chap. 10, Cebesoy depicts preservice teachers' emotive reasoning based on place-based SSI instruction through the example of gold mining in Uşak, Türkiye. This chapter's findings show that place-based instruction has an effect on the frequency and intensity of preservice teachers' emotive reactions, as well as the development of deep compassion, guilt, rage, and righteous indignation.

Chapter 11 provides empirical evidence of how preservice teachers co-design technology-enriched place-based SSI lesson plans. In this chapter, preservice teachers explore water pollution issues in Istanbul, Türkiye, through place-based pedagogies. Then, preservice teachers co-design practices in two place-based SSI topics, namely, biodiversity and water pollution. The chapter concludes that preservice teachers embedded various technology tools for diverse purposes. Also, as new tools were released and made available, preservice teachers' repertoires diversified.

Chapter 12 focuses on preservice teacher education for place-based SSI. Selcan first introduces pedagogical content knowledge for SSI. Then, the chapter shifts our focus to pedagogical content knowledge for place-based SSI. Selcan calls for teacher education courses to enhance preservice teachers' practical knowledge or pedagogical content knowledge on place-based SSI through three practices: providing preservice teachers with effective place-based SSI instruction, making preservice teachers' examine experienced teacher' lesson plans and create lesson structured plans individually or in groups to improve teacher knowledge, and utilizing mentoring practices to improve preservice teachers' place-based SSI instruction.

The final chapter, Chap. 13, combines the main themes discovered through the research conducted in the individual chapters. In this chapter, Bahadir and Engin highlight the following avenues for research and pedagogy: the significance of place that incorporates social and cultural perspectives and dispositions, fostering learner agency, strengthening community engagement, empowering students to take responsible actions through youth activism and civic engagement, fostering environmental stewardship and perspective taking skills, encouraging policymakers to prioritize more place-based issue focused instruction and content in science education, exploring potential cross-cultural perspectives.

Overall, the book project's main contribution is providing science teacher educators with research-based learning activities focusing on SSI. The learning activities address SSI from a variety of contexts throughout Türkiye. Each chapter makes clear connections to cross-cutting issues that go beyond borders, such as character and value enactment or argumentation. Chapters also conclude by transferring the issue to global context sections, where the authors explicitly extend the issue beyond the Turkish context.

This book is primarily intended for educators and researchers in science education, SSI instruction, and place-based pedagogy. The book addresses the needs of educators looking to improve students' meaningful learning and researchers looking to fill gaps in the literature about the intersection of SSI, place-based learning, and other cross-cutting issues like modeling, sense of place, argumentation, character and values, and so on. Overall, the book is a valuable resource for international scholars and educators seeking to engage preservice teachers in meaningful and contextually rich experiences provided by place-based education.

İzmir, Türkiye Bahadır Namdar

Ankara, Türkiye Engin Karahan

References

Birmingham, D., & Calabrese Barton, A. (2014). Putting on a green carnival: Youth taking educated action on socioscientific issues. *Journal of Research in Science Teaching, 51*(3), 286–314.

Calabrese Barton, A., & Tan, E. (2010). We be burnin: Agency, identity and learning in a green energy program. *Journal of Canadian Journal of Science, Mathematics and Technology Education, 10*(3), 207–222.

Caniglia, G., John, B., Bellina, L., Lang, D. J., Wiek, A., Cohmer, S., & Laubichler, M. D. (2018). The glocal curriculum: A model for transnational collaboration in higher education for sustainable development. *Journal of Cleaner Production, 171*, 368–376.

Capkinoglu, E., & Yilmaz, S. (2018). Examining the data component used by seventh grade students in arguments related to local socioscientific issues. *Eğitim ve Bilim-Education and Science, 43*(196), 125–149.

Cook, K., & Buck, G. (2014). Pre-service elementary teachers' experience in a community of practice through a place-based inquiry. *International Journal of Environmental and Science Education, 9*(2), 111–132.

Evren Yapicioglu, A., & Atabey, N. (2020). Evaluation of trends in theses on socio-scientific issues: The case of Turkey. *International Journal of Progressive Education, 16*(4), 115–134.

Gruenewald, D. A. (2003). The best of both worlds: A critical pedagogy of place. *Educational Researcher, 32*(4), 3–12.

Herman, B. C. (2018). Students' environmental NOS views, compassion, intent, and action: Impact of place-based socioscientific issues instruction. *Journal of Research in Science Teaching, 55*(4), 600–638.

Herman, B. C., Olson, J. K., Colbert, J. T., & Holtz, J. D. (2013). The relationship between environmental freechoice learning and students' learning, attitudes, and policy views about waterways. *International Journal of Science and Mathematics Education, 11*(6), 1327–1350. https://doi.org/10.1007/s10763-012-9379-3

Herman, B. C., Owens, D. C., Oertli, R. T., Zangori, L. A., & Newton, M. H. (2019). Exploring the complexity of students' scientific explanations and associated nature of science views within a place-based socioscientific issue context. *Science & Education, 28*, 329–366.

Herman, B. C., Zeidler, D. L., & Newton, M. (2020). Students' emotive reasoning through place-based environmental socioscientific issues. *Research in Science Education, 50*, 2081–2109.

Herman, B. C., Newton, M. H., & Zeidler, D. L. (2021). Impact of place-based socioscientific issues instruction on students' contextualization of socioscientific orientations. *Science Education, 105*(4), 585–627.

Herman, B. C., Poor, S. V., Oertli, R. T., & Schulte, K. (2023). Promoting young learners' NOS views through place-based SSI instruction. *Science & Education, 32*(4), 947–992. https://doi.org/10.1007/s11191-022-00353-z

International Energy Agency. (2021). *Net zero by 2050.* https://www.iea.org/reports/net-zero-by-2050

Karahan, E. (2022). Sosyobilimsel konular ve yer temelli eğitim. [Socioscientific issues and placed based education] B. Namdar (Ed.), *Çağdaş yaklaşımlarla etkinlik destekli sosyobilimsel konuların öğretimi* (ss. 179–199). Nobel Akademi.

Karahan, E., & Roehrig, G. (2016). Use of socioscientific contexts for promoting student agency in environmental science classrooms. *Bartın University Journal of Faculty of Education, 5*(2), 425–442.

Kim, G., Ko, Y., & Lee, H. (2020). The effects of community-based socioscientific issues program (SSI-COMM) on promoting students' sense of place and character as citizens. *International Journal of Science and Mathematics Education, 18*, 399–418.

Lee, H., Yoo, J., Choi, K., Kim, S. W., Krajcik, J., Herman, B. C., & Zeidler, D. L. (2013). Socioscientific issues as a vehicle for promoting character and values for global citizens. *International Journal of Science Education, 35*(12), 2079–2113. https://doi.org/10.1080/09500693.2012.749546

Mathews, J. M. (2013). *Place-based design: An instructional design theory for supporting community-based inquiry and design projects*. Doctoral dissertation. University of Wisconsin-Madison.

McVittie, J., Webber, G., Hellsten, L.-A. M., & Miller, D. (2020). Pathways and pedagogies: Conversations with teacher educators about the role of place-based learning. *Canadian Journal of Environmental Education, 23*(1), 33–49.

Miller, D., & Twum, S. (2017). The experiences of selected teachers in implementing place based education. *In Education, 23*(1), 92–108. https://doi.org/10.37119/ojs2017.v23i1.282

Powell, W. A. (2021). Effects of place-based socioscientific issues on rising middle school students' evidence-based reasoning and critical thinking on hydraulic fracking. *Journal of Education in Science, Environment and Health (JESEH), 7*(3), 220–242. https://doi.org/10.21891/jeseh.961002

Powell, W. A., Newton, M. H., & Zeidler, D. L. (2021). Impact of socioscientific issues on middle school students' character and values for global citizenship. In W. A. Powell (Ed.), *Socioscientific issues-based instruction for scientific literacy development* (pp. 56–91). https://doi.org/10.4018/978-1-7998-4558-4.ch003

Tien, C. Y., & Talley, P. C. (2012). "Think globally, act locally": 'Glocalization' in Taiwanese higher education. *International Journal of Business and Social Science, 3*(15), 124–130.

Topçu, M. S., Yılmaz-Tüzün, Ö., & Sadler, T. D. (2011). Turkish preservice science teachers' informal reasoning regarding socioscientific issues and the factors influencing their informal reasoning. *Journal of Science Teacher Education, 22*(4), 313–332.

Zeidler, D. L., Herman, B. C., & Sadler, T. D. (2019). New directions in socioscientific issues research. *Disciplinary and Interdisciplinary Science Education Research, 1*(1), 1–9.

Contents

1 **Place-Based Socioscientific Issues** 1
Engin Karahan

2 **Developing Preservice Science Teachers' Sense of Place Through Place-Based Socioscientific Issues Instruction** 13
Nilay Ozturk and Büşra Tuncay-Yüksel

3 **Preservice Science Teachers' Socioscientific Orientations Through Placed-Based Socioscientific Issues Teaching: Air Pollution in Ankara** 31
Süleyman Çite and Selçuk Şahingöz

4 **Developing Argumentation Skills in Place-Based SSIs: Construction of a Nuclear Power Plant in Sinop** 45
Nurhan Öztürk and Ayşe Yenilmez Türkoğlu

5 **Preservice Primary School Teachers' Experience of Place-Based Education: Porsuk River** 61
Ahmet Oğuz Akçay and Mehmet Arif Bozan

6 **Preservice Teachers' Enactment of Character and Values Through a Place-Based Drama Activity** 77
Ayşegül Oğuz Namdar, Bahadır Namdar, and Fulden Guler Nalbantoglu

7 **Developing Preservice Science Teachers' Conceptual Understanding Through Place-Based SSI Instruction: Population Growth in Istanbul** 93
Seda Usta Gezer, Çiğdem Çıngıl Barış, and Filiz Avcı

8 **Enhancing Preservice Teachers' Understanding and Application of Place-Based Socioscientific Issues: A Focus on Renewable Energy and Perspective-Taking** 113
Nilay Muslu and Hakan Isik

9 Enhancing Preservice Teachers' Understanding of the Nature of Science through Place-Based Socioscientific Issue Instruction: The Sea-Filled Airport Case 133
Banu Avşar Erümit and Tuğba Yüksel

10 Exploring Preservice Science Teachers' Emotive Reasoning through Place-Based SSI Instruction: The Case of Gold Mining 153
Ümran Betül Cebesoy

11 Codesign of Technology-Enriched Place-Based SSI Lesson Plans . . . 181
Ferah Özer and Çiğdem Han Tosunoğlu

12 Preservice Teacher Education for Place-Based Socioscientific Issues 199
Elif Selcan Öztay

13 New Directions for Place-Based Socioscientific Issue Instruction and Research 215
Bahadır Namdar and Engin Karahan

Editors and Contributors

About the Editors

Bahadır Namdar is an associate professor of science education at Ege University, Izmir, Türkiye. He received his combined B.A. and M.Sc. degrees in biology education from Dokuz Eylül University, Izmir, Türkiye. Dr. Namdar earned his Ph.D. at the University of Georgia in the United States. He worked as a research and teaching assistant in the Department of Mathematics and Science Education at the University of Georgia. He was a visiting student at Iowa State University, Center of Technology in Teaching and Learning, through a Fulbright scholarship. He was also a visiting student at the University of California, Berkeley. His doctoral dissertation focused on the use of multiple representations for socioscientific argumentation in a computer-supported collaborative learning environment. He has been involved in several national and international research projects. His research focuses on argumentation, socioscientific issues, technology integration, and teacher education.

Engin Karahan is an associate professor of science education at Middle East Technical University. After receiving his bachelor's degree from Hacettepe University's Science Education Department, he went on to the University of Minnesota in the United States to pursue his master's and doctorate degrees in STEM education. During his graduate studies, he worked as a researcher at the University of Minnesota's STEM Education Center on NASA and NSF-funded projects that addressed socioscientific issues, culturally relevant climate change education, and educational technologies. His dissertation concentrated on designing and evaluating SSI-focused professional development programs for science teachers. His research areas include STEM education, socioscientific issues, learning technologies, and teacher/student agency.

About the Contributors

Ahmet Oğuz Akçay is an associate professor of primary education at Eskisehir Osmangazi University, Eskisehir, Türkiye. He received his B.A. in primary education from Selcuk University, Konya, Türkiye. He was awarded a full scholarship by the Turkish Ministry of National Education to pursue a master's and doctorate in primary education in the United States. He completed his master's degree in curriculum instruction at the University of Texas at San Antonio in the United States. Dr. Akçay earned his doctorate degree at the Duquesne University, USA, in 2016. Throughout his time in graduate school, he actively engaged in various projects and worked as a researcher primarily focused on teacher professional development of mathematics teaching. His research focuses on mathematics education, STEM education, technology integration, and teacher education.

Filiz Avci is an associate professor in the Department of Mathematics and Science Education at İstanbul University-Cerrahpaşa. She completed undergraduate education in the Department of Chemistry at Istanbul Technical University, in 1999, and her master's degree in the Department of Organic Chemistry at Istanbul University, in 2008. In 2015, she completed her Ph.D. education in the science education doctorate program at Istanbul University. In 2005, she started to work as a Research Assistant in Istanbul University Hasan Ali Yücel Faculty of Education, Department of Primary Education at Science Education program. In 2018, she was appointed as an assistant professor. In 2022, she became a member of Istanbul University-Cerrahpaşa Hasan Ali Yücel Education Faculty, Science Education Department Accreditation Commission. In 2022, she carried out the project titled "The effect of socio-scientific subject-based learning environments on the cognitive structures, emotional tendency and scientific thinking habits of scıence teachers" for the Ministry of National Education. In 2023, she received the title of associate professor in the field of science education. Dr. Avcı works in the fields of science education, chemistry education, cooperative learning, distance learning, and applications of technology in education. She has 12 international and 6 national scientific articles, 50 international and national congress presentations, 17 national projects, 1 book, and 1 book chapter.

Banu Avşar Erümit is an assistant professor at Recep Tayyip Erdogan University in Rize, Türkiye. She began her academic career at Gazi University's Department of Biology Education, where she earned her bachelor's degree. She extended her educational journey by attaining her master's and doctoral degrees in science education at Indiana University, Bloomington, USA, within the Department of Curriculum and Instruction. Additionally, she expanded her academic expertise by securing a minor degree from the Department of Public and Environmental Affairs.

Her doctoral studies provided an opportunity to enhance her academic and professional experiences. During this period, she served as a teaching assistant, delivering various undergraduate courses and actively engaging in numerous educational research projects. Avşar Erümit's research primarily focuses on the nature of science, multiculturalism, and socioscientific issues. These areas of focus have guided her participation in research projects that are both nationally and internationally funded.

Mehmet Arif Bozan completed his bachelor's degree in primary school teaching program at Kırıkkale University, his master's degree in primary education at Eskişehir Osmangazi University, and his doctorate in primary education at Hacettepe University. His master's thesis was on "Professional development of primary school teachers towards STEM education." His doctoral dissertation was on examining "The effect of educational games on refugee students' language learning." He is currently working on refugee students, language teaching, and STEM education. He has taken part in both domestic and foreign funded projects.

Ümran Betül Cebesoy works as an associate professor in the Department of Mathematics and Science Education. She completed her B.A. as Hons in the Department of Science Education at Istanbul University. After completing her M.A. in Science Education Department at Marmara University, she started pursuing her Ph.D. in Elementary Education Department at Middle East Technical University (METU) in 2010. She completed her Ph.D. thesis entitled "An analysis of science teachers' genetics literacy and related decision-making process" in 2014. During her Ph.D., she explored the genetic literacy levels of science teachers and their ability to make decisions on genetics-related socioscientific issues. She visited Stockholm University as a visiting scholar in 2019 and worked with Prof. Dr. Shu-Nu Chang Rundgren on the SEE-SEP model and its use in preservice teacher education programs. She has been working as an associate professor of science education at Usak University since 2018. Her research includes socioscientific issues, decision-making processes, and STEM education.

Çiğdem Çingil Bariş is an associate professor in the Department of Mathematics and Science Education at İstanbul University-Cerrahpaşa. She graduated from the Biology Department of İstanbul University with a bachelor's degree in 2001 and completed her master's program in the Department of Botany at İstanbul University in 2005. In 2012, she completed her Ph.D. in the Department of Botany at İstanbul University. In 2002, she was appointed as a research assistant in the Department of Science Education at the Hasan Ali Yücel Education Faculty, İstanbul University. In 2014, she was appointed as an assistant professor. In 2022, she took on the role of vice chair of the Department of Mathematics and Science Education. In 2023, she received the title of associate professor in the field of biology education. She conducts research in the areas of science education, biology education, and teacher education.

Süleyman Çite is an assistant professor of science education in the Department of Mathematics and Science Education at Kastamonu University. He earned his bachelor degree from the Department of Elementary Science Education at Dokuz Eylül University. Then he completed his master's and doctoral studies in the Learning, Teaching and Curriculum Department at the University of Missouri, Columbia. During his graduate studies, he was involved in various research projects related to improving teacher design capacity to develop and implement innovative curricular resources (SSI units, assessment tasks, inquiry-based science units). Currently, he teaches courses on undergraduate and graduate levels about nature of science, socioscientific issues, and science methods courses. His research interest involves understanding ways to develop both preservice and in-service teachers' practices about designing and implementation of innovative science curriculum materials for improving learners' functional science literacy.

Fulden Guler Nalbantoglu is a full-time research assistant in the Department of Mathematics and Science Education at Ege University. She received her bachelor's degree from Middle East Technical University (METU), Science Education Program, and completed her master's in curriculum and instruction at METU. Dr. Guler-Nalbantoglu earned her Ph.D. in science education at the same university in 2023 with her dissertation entitled "Development of preservice science teachers' pedagogical content knowledge for STEM in the context of lesson study." Her research interests are pedagogical content knowledge, STEM education, and science teacher education.

Çiğdem Han Tosunoğlu is an assistant professor of science education at Marmara University. She graduated from the Biology Education Department of Marmara University with a bachelor's degree in 2008 and completed her master's program in the Department of Biology Education at Marmara University in 2011. She completed her Ph.D. thesis entitled "Examination of biology teachers' pedagogical content knowledge about socioscientific issues" in 2018. During her Ph.D. education, she visited Illinois Institute of Technology as a visiting scholar in 2015 and worked with Prof. Dr. Norman G. Lederman Rundgren. Between 2008 and 2018, she worked as a research assistant at Marmara University, Faculty of Education, Department of Mathematics and Science Education. In 2020, she was appointed as an assistant professor at Marmara University. Her research interests include socioscientific issues, decision-making processes, pedagogical content knowledge, scientific inquiry, and teacher professional development.

Hakan Işik graduated from Çukurova University with a BS in physics. In 2008, he earned his doctorate in science education from Ohio State University. Currently, he is an associate professor of physics education at the Faculty of Education at Muğla Sitki Koçman University, Türkiye. He teaches physics and astronomy courses at the undergraduate level and inquiry-based science courses at the graduate level. His area of interest is creating inquiry-based physics teaching and learning methods.

Nilay Muslu received a combined bachelor's and master's degree in physics education from Dokuz Eylul University. Subsequently, she was awarded a full scholarship by the Turkish Ministry of National Education to pursue a master's and Ph.D. in science education in the United States. She received her master's and Ph.D. degrees from the University of Missouri-Columbia, in learning, teaching, and curriculum with an emphasis in science education. Throughout her time in graduate school, she actively engaged in various projects, primarily focused on teacher professional development and classroom assessment practices. Furthermore, she played integral roles as a researcher and instructor in several projects supported by the Scientific and Technological Research Council of Türkiye (TÜBİTAK). Currently, Dr. Muslu serves as an instructor and researcher at the Muğla Sıtkı Koçman University, specifically within the Faculty of Education, Department of Mathematics and Science Education, and Department of Science Education. Her research interests encompass a wide range of areas, including teacher education, formative assessment, physics education, and technology education.

Ayşegül Oğuz Namdar graduated from Hacettepe University, Faculty of Education, Primary Education Department, Primary School Teaching. She completed her graduate education in Hacettepe University Primary Education Department in 2009, and her doctorate education in 2013 in the same university and department. She wrote her thesis titled "Comparison of art lesson curriculum of 1992 with visual arts curriculum of 2006" at the master's level and "The effect of creative drama method on primary school teacher candidates' interest in art and art education lesson" at the doctoral level. She has articles, conference papers, and book chapters at the national and international level in the fields of creative drama and art education. She worked as a director, researcher, and trainer in many projects and scientific events supported by TÜBİTAK, ERASMUS, and Office of Scientific Research Projects. She has advised master's theses on creative drama, plastic arts, and arts education. She has taught drama lessons in several universities abroad as part of faculty exchange programs. She has been working as a faculty member at Recep Tayyip Erdoğan University, Faculty of Education, Department of Primary Education, since 2009. She has received the title of associate professor in primary education/art in 2017. She still conducts compulsory/elective drama courses at the departments under the Faculty of Education.

Ferah Özer received her B.A., M.A., and Ph.D. degrees from Abant İzzet Baysal University, Türkiye. During her undergraduate and graduate studies, she studied in Sweden and in Spain with exchange programs. During her master's education, she worked as a scholar in the nationwide TUBITAK 1001 R&D project titled "Continuing Teacher PD to Support the Teaching About NOS." Her Ph.D. thesis project entitled "The investigation of the impact of the science learning environments promoting 21st century skills' development on middle school students' problem-solving skills, creative thinking abilities, and conceptual learning" was supported by the Research Projects Unit of the University as well as was granted by the U.S. Department of State-Fulbright Program under the Ph.D. Thesis Research

Fellowship. She conducted part of her research as a visiting scholar under the supervision of Fouad Abd-El Khalick at the University of North Carolina at Chapel Hill. Dr. Özer worked as a research assistant at Boğaziçi University and later taught undergraduate courses as an instructor at the same university. Dr. Özer currently works at Koç University as a learning experience designer and an instructor, lecturing graduate students and advising faculty members. Her research interests include problem solving-creative thinking skills, nature of scientific inquiry, and teacher professional development programs.

Elif Selcan Öztay is an associate professor of science education in the Department of Mathematics and Science Education at Van Yüzüncü Yil University, Türkiye. She graduated from the Chemistry Education Department of Gazi University with combined B.A. and M.Sc. degrees in 2007. In 2016, she received her Ph.D. from the Department of Secondary Science and Mathematics Education at Middle East Technical University (METU) with her Ph.D. thesis entitled "Examination of interaction between pre-service chemistry teachers' pedagogical content knowledge and content knowledge in electrochemistry." During her doctoral education, with the acceptance of Prof. Dr. Julie Anne Luft, she worked as a visiting scholar in the 2012–2013 academic year in the Department of Mathematics and Science Education at the University of Georgia on developing preservice and in-service science teachers' pedagogical content knowledge. Her research interests include integrated STEM education, science education, science teacher education, and pedagogical content knowledge.

Nilay Ozturk is an associate professor at the Faculty of Educational Sciences, Bahcesehir University, and BAUSTEM Center. She received her B.A., M.S., and Ph.D. degrees from Middle East Technical University (METU) in Elementary Science Education, Elementary Science and Mathematics Education, and Elementary Education, respectively. For a period of 6 months during her Ph.D. studies, she was awarded the Scientific and Technological Research Council of Türkiye Scholarship to work with Dr. Troy Sadler at the University of Missouri. After completing Ph.D., she worked with Dr. Gillian H. Roehrig at the University of Minnesota for 1 year to complete her postdoctoral research project funded by the Scientific and Technological Research Council of Türkiye Scholarship Program. During the postdoctoral research project in Minnesota, she worked with middle school teachers from different disciplines to design and implement integrated STEM curricular units in the context of locally relevant socioscientific issues. Her research interests focus on socioscientific issues-based education, integrated STEM education, science education, and teacher education.

Nurhan Öztürk works as an associate professor in the Department of Mathematics and Science Education. In 2006, she graduated in Science Education program from Kocaeli University. She completed her master's degree at Eskişehir Osmangazi University in 2008 and her doctoral degree at Gazi University in 2013. Between 2011 and 2013, she worked as a research assistant at Sinop University, Faculty of

Education, Department of Mathematics and Science Education. In 2020, she was appointed as associate professor and is still working as an associate professor at Sinop University. She carried out several activity-based TUBITAK and EU-supported projects in the fields of socioscientific issues, argumentation, STEM, and out-of-school learning environments and has taken part in many other projects. She published several scientific studies on these topics and served as a supervisor for several master's thesis studies.

Selçuk Şahingöz is an assistant professor in the Department of Educational Sciences at Gazi University, Türkiye. He earned his B.S. from Gazi University. Then, he was awarded a scholarship from the Republic of Türkiye the Ministry of National Education and completed his M.S. and Ph.D. degrees at Mallinson Institute for Science Education, Western Michigan University, USA. He worked in the development process of the formative assessment instrument which is entitled Pedagogy of Science Teaching Test (POSTT). Dr. Şahingöz teaches different types of educational courses regarding preservice teacher education, teacher professional development, and curriculum development in science education. His research broadly focuses on inquiry-based learning, socioscientific issues, science teaching pedagogical orientations, and humor-based STEM education. He is also interested in research on twenty-first-century thinking skills in education, especially critical thinking. Dr. Şahingöz is an active member of the National Science Teachers Association (NSTA) and National Association of Research in Science Teaching (NARST).

Büşra Tuncay-Yüksel is an assistant professor of science education in Giresun University, Türkiye. She completed her undergraduate education (major) in the Department of Science Education at Middle East Technical University (METU) in 2007 and received her minor degree from the Biology Department of the same university in the same year. In 2010, she received her Master of Science degree in the Department of Elementary Science and Mathematics Education in METU with her master's thesis entitled "Moral reasoning of pre-service science teachers toward local and non-local environmental problems." During her doctoral education, with the acceptance of Prof. Dr. Dana L. Zeidler, she worked as a visiting scholar for 1 year in the University of South Florida. In 2016, she defended her dissertation entitled "Environmental moral reasoning patterns of pre-service science teachers and their relationships to epistemological beliefs and values" and completed her doctoral studies in the Elementary Education Department in METU. Her academic interests include environmental moral reasoning, values, epistemological beliefs, nature of science, socioscientific issues, and informal learning environments.

Seda Usta Gezer is an assistant professor at İstanbul University-Cerrahpaşa Hasan Ali Yücel Faculty of Education, Department of Mathematics and Science Education. She graduated from Istanbul University, Faculty of Science, Biology Department Undergraduate Program in 2005 and from Istanbul University, Institute of Science, Botany Department Master's Program in 2008. In 2014, she completed Science Education Program in Primary Education Department at the

Institute of Educational Sciences, Marmara University, and received her doctorate. She was appointed as a research assistant in Science Education Department, Hasan Ali Yücel Education Faculty, Istanbul University, in 2006 and as an assistant professor in 2017. In 2011–2012 academic year she was a visiting scholar in the Department of Mathematics and Science Education at the University of Georgia, USA, working with Prof. Dr. J. Steve Oliver. Dr. Gezer is involved as the assistant coordinator in the "Istanbul STEM Learning Center" Project supported by the Istanbul Development Agency, Children and Youth Financial Support Program. She served as a project executive and trainer in various TÜBİTAK 4004 projects and as a researcher in the Ministry of National Education projects. Her main research interest areas are science education, reflective thinking, inquiry-based learning, out-of-school learning, and STEM.

Ayşe Yenilmez Türkoğlu is associate professor for science education at Alanya Alaaddin Keykubat University. She completed her Bachelor's degree in the Department of Science Education at Middle East Technical University (METU) in 2003. In 2006, she received her Master of Science degree in the Department of Elementary Science and Mathematics Education at METU with her master's thesis entitled "Exploring relationships among students' prior knowledge, meaningful learning orientation, reasoning ability, mode of instruction and understanding of photosynthesis and respiration in plants." During her doctoral education, she studied as a doctoral student in Hamburg University between 2010 and 2011. In 2013, she received her Ph.D. degree in the Department of Elementary Education at METU with her dissertation entitled "Investigating pre-service science teachers' understandings and use of models." She worked as a research assistant throughout her graduate education at METU. Dr. Yenilmez Turkoglu currently teaches several graduate and undergraduate courses and supervises graduate theses, and her research focuses on socioscientific issues, models and modeling, conceptual learning, and nature of science and technology.

Tuğba Yüksel is an associate professor in the Science Education Department at Recep Tayyip Erdogan University in Türkiye. She earned her doctoral degree in the Curriculum and Instruction Department at Purdue University in the United States. Throughout her doctoral studies at Purdue University, she served as a research assistant on projects focusing on curriculum development for K-12 science and advanced physics courses, which allowed her to specialize in this area. Dr. Yuksel's research interests revolve around examining how individuals construct and utilize meta models to foster a meaningful understanding of science. Additionally, she is involved in designing model-based learning environments supported by physical, mathematical, and virtual manipulatives. Her research extends to studying active learning in physics education, learning experiences in out-of-school environments, and STEM education. Furthermore, Dr. Yuksel maintains an ongoing collaboration with the Purdue University Cyberlearning Consortium, where she serves as an international partner.

Contributors

Ahmet Oğuz Akçay Department of Primary School Education, Faculty of Education, Eskişehir Osmangazi University, Eskişehir, Türkiye

Filiz Avci Department of Mathematics and Science Education, Hasan Ali Yücel Education Faculty, İstanbul University-Cerrahpaşa, İstanbul, Türkiye

Banu Avşar Erümit Department of Mathematics and Science Education, Faculty of Education, Recep Tayyip Erdoğan University, Rize, Türkiye

Mehmet Arif Bozan Department of Primary School Education, Faculty of Education, Istanbul Aydın University, İstanbul, Türkiye

Ümran Betül Cebesoy Department of Mathematics and Science Education, Faculty of Education, Usak University, Usak, Türkiye

Çiğdem Çingil Bariş Department of Mathematics and Science Education, Hasan Ali Yücel Education Faculty, İstanbul University-Cerrahpaşa, İstanbul, Türkiye

Süleyman Çite Department of Mathematics and Science Education, Faculty of Education, Kastamonu University, Kastamonu, Türkiye

Fulden Guler Nalbantoglu Department of Mathematics and Science Education, Faculty of Education, Ege University, Izmir, Türkiye

Çiğdem Han Tosunoğlu Department of Mathematics and Science Education, Faculty of Education, Marmara University, Istanbul, Türkiye

Hakan Işik Department of Science Education, Faculty of Education, Muğla Sıtkı Koçman University, Muğla, Türkiye

Engin Karahan Department of Mathematics and Science Education, Faculty of Education, Middle East Technical University, Ankara, Türkiye

Nilay Muslu Department of Science Education, Faculty of Education, Muğla Sıtkı Koçman University, Muğla, Türkiye

Bahadır Namdar Department of Mathematics and Science Education, Faculty of Education, Ege University, Izmir, Türkiye

Ayşegül Oğuz Namdar Department of Primary Education, Faculty of Education, Recep Tayyip Erdoğan University, Rize, Türkiye

Ferah Özer Office of Learning & Teaching (KOLT), Koç University, Istanbul, Türkiye

Elif Selcan Öztay Department of Mathematics and Science Education, Faculty of Education, Van Yüzüncü Yil University, Van, Türkiye

Nilay Ozturk Department of Primary Education, Faculty of Educational Sciences & BAUSTEM Center, Bahcesehir University, Istanbul, Türkiye

Nurhan Öztürk Department of Mathematics and Science Education, Faculty of Education, Sinop University, Sinop, Türkiye

Selçuk Şahingöz Department of Mathematics and Science Education, Faculty of Education, Gazi University, Ankara, Türkiye

Büşra Tuncay Yüksel Department of Mathematics and Science Education, Faculty of Education, Giresun University, Giresun, Türkiye

Seda Usta Gezer Department of Mathematics and Science Education, Hasan Ali Yücel Education Faculty, İstanbul University-Cerrahpaşa, İstanbul, Türkiye

Ayşe Yenilmez Türkoğlu Department of Mathematics and Science Education, Faculty of Education, Alanya Alaaddin Keykubat University, Alanya, Antalya, Türkiye

Tuğba Yüksel Department of Mathematics and Science Education, Faculty of Education, Recep Tayyip Erdoğan University, Rize, Türkiye

Chapter 1
Place-Based Socioscientific Issues

Engin Karahan

1.1 Origins

Science education systematically makes sense of the events, changes, and substances that occur around us while also allowing us to get to know ourselves and, as a result, assisting us in making sense of the world (Yuzuak & Zihni, 2022). Learners' lived experiences are frequently overlooked in schools, resulting in cognitive dissonance. School science content is frequently criticized for being decontextualized from real-life situations. Traditional science education promotes a universalist value system encompassing all knowledge and transcending cultural, linguistic, and regional differences while ignoring other value systems to understand the world and explain natural phenomena (Li & Shein, 2022). This viewpoint leads to a conflict between learning theories and the way science is taught in formal school settings. Worksheets and busywork are insufficient for meaningful learning; authentic, problem-based, community-centered, and student-centered activities are required (Arthur, 2022). Hence, school science education must be restructured to emphasize the importance of the context in which learners are situated.

Because of scientific and technological advancements, each period, with its own occurrences, necessitates different teaching and learning processes based on its needs and characteristics. Despite these educational changes, some educational approaches have remained constant. Place-based education is one of those approaches that has been around for a long time. While some believe place-based education is a contemporary approach, others argue that its beginning can be traced

---

E. Karahan (✉)
Department of Mathematics and Science Education, Faculty of Education,
Middle East Technical University, Ankara, Türkiye
e-mail: enginka@metu.edu.tr

B. Namdar, E. Karahan (eds.), *Socioscientific Issues Focused Teacher Education*, Contemporary Trends and Issues in Science Education 60,
https://doi.org/10.1007/978-3-031-55233-5_1

back to ancient times. Place-based education has a well-established historical tradition of learning. The history of place-based education can be traced back to environmental education, which placed emphasis on the sciences, conservation, and sustainable development (Sobel, 2013). Children gain knowledge of the world around them prior to formal education through play, observation, and conversation with their social environment (Smith, 2016).

According to most, the origins of place-based education can be traced back to Dewey, who argued that the school was life itself rather than a place that prepared students for life. Elbaz (2023) described Dewey's educational philosophy as follows:

> At the end of the nineteenth century in the United States, John Dewey gradually included knowledge of the place as a basic part of his curriculum as he sought to overcome the separation between the school and the children's world in a school. He sees that school wastage comes from the inability to benefit from the experiences that learners get outside the school in a complete and meaningful way, as the learners are not able to apply what they have learned in school in their lives. (p. 60)

John Dewey strongly opposed education behind artificial walls and advocated for effective, contextualized learning in the real world. As a result, it is reasonable to assert that the origins of place-based education can be traced back to Dewey. Boxton and Provenzo Jr. (2011), however, asserted that nineteenth-century educators like Pestalozzi and Froebel incorporated place-based learning activities and models into education and emphasized the importance of studying nature, horticulture, and related topics.

1.2 Place-Based Education

Before describing place-based education, it is essential to understand the idea of place. The word "place" is often used to describe any specific inhabited area or geographical location. However, the term "place" can also be used to refer to more nebulous ideas, such as a specific location, one's standing in a social or cultural group, or an area with deep emotional significance. In place-based education, "place" is more than just a geographical context or signifier. Waite (2013) argues that "knowing" a place entails showing an appreciation for its complexities and being aware of how it compares to and differs from other places. Perceptual, sociological, ideological, political, and ecological are the five 'dimensions of place' proposed by Gruenewald (2003). These factors do not operate in isolation; rather, they overlap to create a more complete picture of a location.

The idea of "place-based education" transcends traditional academic disciplines. As a result, various definitions have been proposed, each with its own set of assumptions and limitations. Place-based education is a method of instructing students in all subject areas through exposure to their immediate surroundings. Put another way, it emphasizes hands-on, real-world learning experiences through the use of the local community and environment to teach concepts across the curriculum (Sobel, 2013). To foster a greater connection to local contexts, cultures, and environments, place-based education is defined as an umbrella term for pedagogical practices that prioritize experiential, community-based, and contextual learning (Gruenewald, 2003).

When comparing the various resources that define place-based education, it becomes clear that there are shared characteristics, such as the value placed on local contexts and authentic learning opportunities. Although place-based education originates in environmental education, it can be distinguished from more traditional forms of environmental education due to its proponents' emphasis on the interplay between different environments (Smith, 2007). Place-based education is an approach that considers both the social and natural environment; in other words, place-based education is a teaching and learning strategy that incorporates the surrounding community. The primary goal of place-based education is to empower students to make positive contributions to their communities and the world at large.

The five pillars of place-based education, as identified by McClennen (2016), are as follows: "(1) Connection and relevance, (2) partnerships and permeability, (3) inquiry and design, (4) student-centered, and (5) interdisciplinary." Each characteristic is described below:

- One of the most important aspects of a successful education is effective communication with the communities where students live and work.
- Strong alliances between various educational institutions and society partners are essential to place-based education.
- Research, inquiry, and design are all crucial parts of a place-based education that encourages students to gain insight into the world around them by posing questions, articulating problems, developing hypotheses, amassing data, and developing solutions.
- Place-based education is learner-centric, meaning that it relies on the student's participation and initiative throughout the course of study for the best results.
- Place-based education is not limited to a single academic field but rather encompasses all fields of inquiry that can contribute to an understanding of the location.

Nature studies, cultural studies, induction into societal processes, real-world problem-solving, and training and entrepreneurial opportunities are among the five thematic models in place-based education (Smith, 2002). As previously stated, there are various approaches to place-based education. Despite the differences in place-based education applications, there are some common features that must be addressed. Smith (2002) listed five characteristics shared by place-based applications. First, lessons are tailored to the community's needs and resources. Second, rather than receiving information passively, students who engage with their surroundings actively construct their own understanding. Third, students cultivate an interest in and ability to investigate and solve problems facing their immediate surroundings. Fourth, educators encourage and facilitate students' participation in local studies. Fifth, the school's ties to the neighborhood strengthen.

According to Granit-Dgani (2021), place-based education has four main components. The following is a summary of these features (Granit-Dgani, 2021, cited in Yemini et al., 2023) (Fig. 1.1).

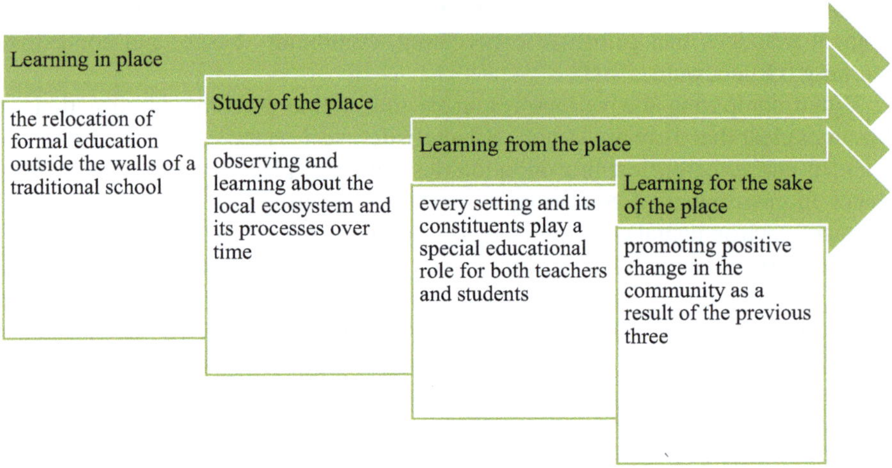

Fig. 1.1 Main components of place-based education

Using a place-based education also has a number of benefits for learners. Place-based education is an emerging strategy that brings new advantages to the classroom by setting lessons in the context of students' immediate surroundings (Chinn, 2015). In addition to strengthening students' connection to their local communities, place-based education has been shown to increase their critical thinking abilities and engagement in class (Li & Shein, 2022). Kuwahara (2010) argues that the primary goal of place-based education is to empower students to make positive contributions to their communities and the world. The benefits of place-based education extend beyond the individual student to the larger community. Consequently, it is linked to other pedagogical strategies, such as eco-ed and sustainability-ed. In this way, they carry out tasks that achieve a better future for their society (Kuwahara, 2010).

1.3 Sense of Place

Having a "sense of place" is often cited as a vital component of place-based education. A person's or a group's sense of place is their personal and emotional connection to a physical location. Space and place are fundamental aspects of living in the world, and a sense of place is formed through personal experiences (Li & Shein, 2022). Relationships between people and their surroundings are the source of their sense of identity and belonging. The physical, cultural, historical, and social features that make up a place's distinct character and atmosphere are all included in what we call its "sense of place." One's own history, recollections, and interactions with their surroundings are major contributors to their sense of place. It can be affected by things like topography, buildings, landmarks, customs, language, social interactions, and general atmosphere.

The goal of place-based education is to boost students' sense of place by encouraging them to use what they have learned to address issues in their own backyards (Kosker & Karabag, 2012). The community itself serves as a learning context in place-based education, and students' work is oriented toward community priorities, with members of the community serving as both partners and resources in the classroom (Uztemur & Dere, 2022). Community involvement and the acquisition of skills for social integration are fostered by a regional emphasis (Smith & Sobel, 2010).

Although the term "place" is often misused in place-based education, "place" refers to more than just a physical or geographical location. Cultural commons, for example, are frequently emphasized in place-based education research and practice. As a result, students' sense of place is expected to be enhanced by social and cultural enterprises. Considering the significance of a place for meaningful learning, it is crucial to acknowledge the dispositions of local communities, such as attitudes, knowledge, habits, and skills. Ardoin (2006) offers a model to define "place" for the purposes of place-based education. The biophysical refers to the material conditions of a given location; the psychological to each person's unique experience there; the sociocultural to how one is embedded in a particular culture that shapes their perception of and connection to a given location; and the political-economic to how power dynamics shape both the built environment and its inhabitants.

As seen from the aforementioned studies, place-based education relies heavily on the participation of the local community. Therefore, place-based education is predicated on the idea that local communities hold knowledge, skills, and mutual-support systems that can mitigate the negative effects of the consumer culture that is now pervasive in our global society (Bowers, 2008). As stated by the North American Association for Environmental Education, the goal of place-based education is to help students understand the environmental issues related to their lives in such a way that local issues are given priority over global issues (Li & Shein, 2022). Furthermore, place-based education instructs students on how to interact with a specific location in order to fortify their sense of belonging there. The importance of local context cannot be overstated because it not only provides students with a deeper understanding of the region's history, environment, and culture but also strengthens their emotional connection to the area, which in turn encourages them to take better care of it (Sobel, 2004).

According to Li and Shein (2022), the sense of place can be roughly divided into cognitive, socio-emotional, and behavioral, mirroring the educational theory proposed by Bloom. The following table defines the cognitive, socio-emotional, and behavioral dimensions of a sense of place and provides examples of each (Table 1.1).

Table 1.1 Three dimensions of sense of place (Li & Shein, 2022, pp. 5–6)

Cognitive knowledge	Socio-emotional knowledge	Behavioral knowledge
General living environment: The ability to understand the differences between the local environment and other places Specific place characteristics: The ability to provide direct descriptions of the environment Geography: The ability to compare differences among various places and integrate geographical characteristics into systematic knowledge Culture: Understanding of the local culture Cultural meanings: The understanding of the meanings behind geography and culture Value judgment: The ability to contemplate the changes in geography and culture under different spatiotemporal contexts	Recognizing a place: Producing an environmental image composed of people, matters, things, and related memories, but only for identification without other meanings Belonging to a place: An individual connects with the place and finds significance in local affairs Identifying with the place goals: Defining self-identity and status based on ideals, beliefs, preferences, feelings, values, goals, behavioral tendencies, and intricate patterns of skills that are related to the place Depending on a place: An individual evaluates the meaning of the place to oneself and generates a strong psychological connection with the place Involving in a place: Actively investing in local affairs with practical actions Rooting in a place: a strong sense of connection, security, and comfort with the place, leading to solid motivations for making a commitment to the place and being willing to sacrifice personal interests for the place	Observation: Finding special things in the environment to observe Identifying problems: Finding problems that are worth exploring or need to be resolved from everyday life Planning: Thinking about the identified problems and planning a feasible solution Systems thinking: Analyzing information and resources from all sources and considering a holistic view of the problems or issues Acting: Starting to put ideas into practice Problem-solving: Helping local people solve problems in life

1.4 Socioscientific Issues

The public is constantly exposed to novel dangers in the twenty-first century due to the widespread application of science and technology. Therefore, Beck (1992) uses the term "risk society" to characterize a contemporary industrial society. Dangers associated with the growth-oriented paradigm in society include things like sinkholes and the collapse of buildings and bridges, while the double-edged sword of science and technology includes things like genetically modified organisms, embryonic stem cell research, and artificial additives in food (Kim et al., 2020). Science education should equip students to deal with those ill-structured cases so that societies can have informed citizens.

Over the past few decades, increasing students' scientific literacy has emerged as a central goal of science curricula around the world. Achieving scientific literacy for all students has been a consistent focus of reform movements in science education (Bybee, 1997). The concept of scientific literacy has been the subject of ongoing debate among science educators (DeBoer, 2000; Hurd, 1998). The goal of creating scientifically literate citizens has prompted a number of initiatives in science education. The incorporation of SSI into the classroom is one approach to improving students' understanding of the practical applications of science (Zeidler et al., 2005). Students need to be able to navigate SSI (Sadler et al., 2004) related to the applications of science and technology before they can be considered scientifically literate citizens. The primary goal of this SSI framework is to cultivate individuals who are well-rounded in terms of their knowledge, skills, and values in order to make sound decisions (Zeidler et al., 2005).

SSI are contentious and complex scientific issues with social consequences (Herman, 2018). They are woven into the cultural, political, and moral fabric of our lives, affecting our communities on a personal, regional, and global scale (Lee et al., 2013). Therefore, resolving them using only scientific knowledge and perspectives may be hampered. When it comes to morality and ethics, SSI are often at the center of heated debate (Zeidler & Kahn, 2014). Climate change, GMO foods, the effects of agriculture on water supplies, and the introduction or reintroduction of species are just a few of the many examples of SSI (Sadler & Murakami, 2014). In order to address the outcomes of scientific literacy, various curricula around the world include objectives related to SSI.

The literature suggests that there are numerous pedagogical affordances of SSI. This is especially true for academic interventions that are relevant to everyday life, which is why using SSI as a context for teaching and learning helps students understand and apply fundamental science concepts (Herman et al., 2020). However, it is crucial to use the SSI framework effectively when creating educational processes. According to Herman et al. (2021), essential design elements for effective and impactful SSI instruction include organizing instruction around a compelling issue, foreshadowing the issue early, scaffolding students' understanding through higher-order practices, and engaging students in a culminating experience. Another

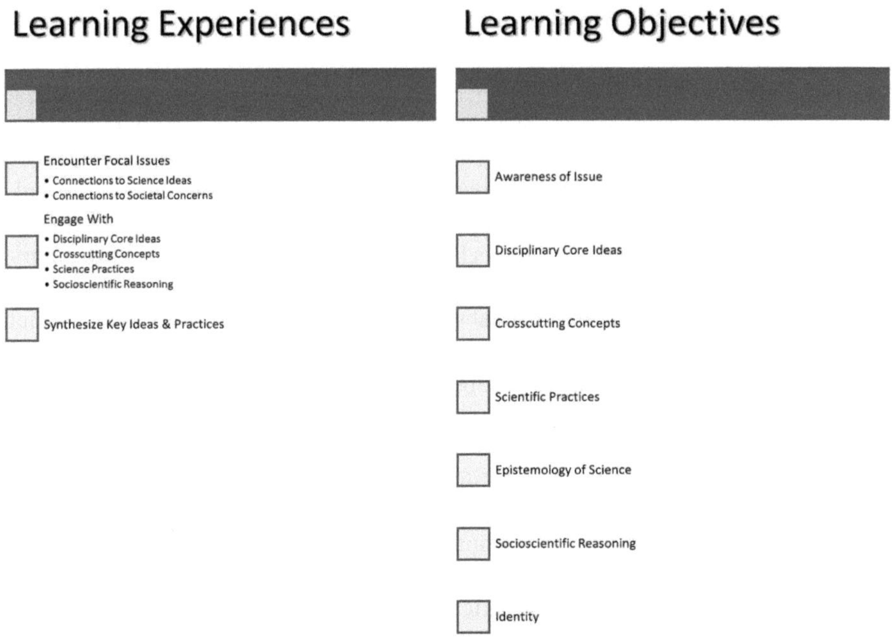

Fig. 1.2 Graphic representation of the SSI teaching and learning model (Sadler et al., 2017, p. 80)

desired design element is the use of media to connect learning activities to the "real world" (p. 589).

The main objective of teaching SSI is to include students in the scientific process by having them investigate topics that are important to them on a personal level as well as to society and the world at large (Zeidler, 2015). Understanding how to teach SSI more effectively is essential for achieving these objectives. In their work, Sadler et al. (2017) described the development and evolution of a model for SSI teaching and learning (SSI-TL). Based on the model, SSI-TL should incorporate a specific set of learning experiences with specific goals for student achievement (see Fig. 1.2).

The structure of the model is described as follows (Sadler et al., 2017):

> The left side of the model presents the sequence of learning experiences. This sequence is made up by three main phases. The first phase involves students encountering the focal issue. The second phase captures the main body of teaching and learning experiences and calls for student engagement with science ideas, science practices, and socioscientific reasoning practices. We conceptualize the learning of science consistent with NGSS and its call for three-dimensional science learning, including emphases on disciplinary core ideas, crosscutting concepts, and scientific practices. The third phase of the model corresponds with a culminating experience in which students synthesize the ideas and practices with which they have engaged throughout the unit. (p. 78)

Certain results have been recorded in the literature as a result of empirical investigations into SSI education. Some examples of outcomes are as follows (Zeidler, 2015, para. 10):

- Promoting developmental changes in reflective judgment
- Moving students to more informed views of the nature of science
- Increasing moral sensitivity and empathy
- Increasing conceptual understanding of scientific content
- Increasing students' ability to transfer concepts and scaffold ideas
- Revealing and reconstructing alternative perceptions of science
- Facilitating moral reasoning
- Improving argumentation skills
- Promoting understanding of eco-justice and environmental awareness
- Engaging students' interest in the inquiry of science

Note that learners should be given meaningful experiences that are relevant to their lives for SSI pedagogy to reach its full potential. Thus, the SSI context should incorporate sociocultural elements of places where students' experiences are shaped. So, in what follows, we will discuss how SSI can be taught and learned through the use of place-based pedagogies.

1.5 Place-Based Socioscientific Issues

As previously mentioned, the content and practices of science education are often divorced from their real-life contexts. Authentic engagement with the sociocultural, moral, and affective perspectives of SSI is often overlooked in favor of a focus on discrete learning goals in most formal science learning experiences (Herman, 2018). Many educators have voiced their disapproval of this method, advocating instead for more "real-world" situations in the classroom. Contextual learning takes place in the learners' everyday interactions with the people and environments they spend the most time in. Students can develop a stronger sense of belonging to their local community and a sense of ownership over addressing and resolving SSI through the use of place as a context for learning (Stevenson et al., 2014; Tytler et al., 2017). That is why it is so important to make use of the pedagogical possibilities afforded by place-based education when dealing with SSI in the classroom.

Since science classes "can distance students from the lived SSI challenges their communities face and often fail to position youth as contributors and participants in the pursuit of possible and desperately needed solutions" (Bang & Marin, 2015, p. 531), it seems imperative to use place-based SSI educational approaches in order to foster desired and intertwined nature-culture relations among students (Herman et al., 2020). Fostering students' sense of place and attachment to nature through a synthesis of experiential and instructional environmental education approaches in meaningful contexts is emphasized by Kudryavtsev et al. (2012). Together, these strategies would give students numerous opportunities to interact directly with people and ecosystems that have been affected by SSI.

According to studies, teaching about place-based SSI helps students form deep emotional ties to the natural world, which in turn raises their level of environmental consciousness and their sense of agency (Herman et al., 2020). This is exactly why it is important to employ SSI-based learning opportunities that are grounded in place-based contexts. There is growing recognition that place-based education is an effective method for educating future educators (Dani, 2019). Using place-based tools, aspiring educators can gain a deeper comprehension of the cultural and social aspects of the classroom. Prospective educators benefit from these routines because they foster a sense of belonging and enhance their capacity to create meaningful learning activities (Uztemur & Dere, 2022). In order for future teachers to understand the social and cultural context in which their students will be working, it is crucial for teacher educators to transform communities into learning labs (Buxton & Provenzo Jr., 2011). Given that place-based SSI provides rich and meaningful learning experiences for students, researchers should explore how instruction in place-based SSI occurs in teacher education programs.

References

Ardoin, N. M. (2006). Toward an interdisciplinary understanding of place: Lessons for environmental education. *Canadian Journal of Environmental Education (CJEE), 11*(1), 112–126.

Arthur, M. (2022). Exploring place-based education while reflecting on my own place growing up in West Virginia. *Impacting Education: Journal on Transforming Professional Practice, 7*(3), 40–45.

Bang, M., & Marin, A. (2015). Nature–culture constructs in science learning: Human/non-human agency and intentionality. *Journal of Research in Science Teaching, 52*(4), 530–544.

Beck, U. (1992). *Risk Society*. Sage.

Bowers, C. (2008). Why a critical pedagogy of place is an oxymoron. *Environmental Education Research, 14*(3), 325–335.

Buxton, C. A., & Provenzo, E. F., Jr. (2011). *Place-based science teaching and learning: 40 activities for K-8 classrooms*. Sage.

Bybee, R. W. (1997). *Achieving scientific literacy: From purposes to practices*. Heinemann.

Chinn, P. W. (2015). Place and culture-based professional development: Cross-hybrid learning and the construction of ecological mindfulness. *Cultural Studies of Science Education, 10*, 121–134.

Dani, D. (2019). A community and place-based approach to middle childhood science teacher education. *Middle School Journal, 50*(2), 45–52.

DeBoer, G. E. (2000). Scientific literacy: Another look at its historical and contemporary meanings and its relationship to science education reform. *Journal of Research in Science Teaching, 37*(6), 582–601.

Elbaz, M. M. (2023). Place-based education: Community as a multidisciplinary learning environment. *Port Said Journal of Educational Research, 2*(1), 59–74.

Gruenewald, D. A. (2003). Foundations of place: A multidisciplinary framework for place-conscious education. *American Educational Research Journal, 40*(3), 619–654.

Herman, B. C. (2018). Students' environmental NOS views, compassion, intent, and action: Impact of place-based socioscientific issues instruction. *Journal of Research in Science Teaching, 55*(4), 600–638.

Herman, B. C., Zeidler, D. L., & Newton, M. (2020). Students' emotive reasoning through place-based environmental socioscientific issues. *Research in Science Education, 50*, 2081–2109.

Herman, B. C., Newton, M. H., & Zeidler, D. L. (2021). Impact of place-based socioscientific issues instruction on students' contextualization of socioscientific orientations. *Science Education, 105*(4), 585–627.
Hurd, P. D. (1998). Scientific literacy: New minds for a changing world. *Science Education, 82*(3), 407–416.
Kim, G., Ko, Y., & Lee, H. (2020). The effects of community-based socioscientific issues program (SSI-COMM) on promoting students' sense of place and character as citizens. *International Journal of Science and Mathematics Education, 18*, 399–418.
Köşker, N., & Karabağ, S. (2012). Coğrafya eğitiminde yer temelli öğretim yaklaşımına ilişkin öğretmen görüşleri [Teacher views on place based education approach in the education of geography]. *Türkiye Sosyal Araştırmalar Dergisi, 163*(163), 123–137.
Kudryavtsev, A., Stedman, R. C., & Krasny, M. E. (2012). Sense of place in environmental education. *Environmental Education Research, 18*(2), 229–250.
Kuwahara, J. H. (2010). *Effectiveness of place-based science curriculum projects situated in Hawaiian and Western cultural institutions at an urban high school in Hawai'i. Unpublished doctoral dissertation*. University of Hawai'i at Manoa.
Lee, H., Kyunghee, C., Kim, S., Jungsook, Y., Krajcik, J. S., Herman, B. C., & Zeidler, D. L. (2013). Socioscientific issues as a vehicle for promoting character and values as global citizens. *International Journal of Science Education, 35*(12), 2079–2113.
Li, W. T., & Shein, P. P. (2022). Developing sense of place through a place-based indigenous education for sustainable development curriculum. *Environmental Education Research, 29*(5), 692–714.
McClennen, N. (2016, May 9). *Place-based education: Communities as learning environments*. Getting Smart. https://www.gettingsmart.com/2016/05/09/place-based-education-communities-as-learning-environments/
Sadler, T. D., & Murakami, C. D. (2014). Socio-scientific issues-based teaching and learning: Hydrofracturing as an illustrative context of a framework for implementation and research. *Brazilian Journal of Research in Science Education, 14*(2), 331–342.
Sadler, T. D., Chambers, F. W., & Zeidler, D. L. (2004). Student conceptualizations of the nature of science in response to a socioscientific issue. *International Journal of Science Education, 26*(4), 387–409.
Sadler, T. D., Foulk, J. A., & Friedrichsen, P. J. (2017). Evolution of a model for socio-scientific issue teaching and learning. *International Journal of Education in Mathematics, Science and Technology, 5*(2), 75–87.
Smith, G. A. (2002). Place-based education: Learning to be where we are. *Phi Delta Kappan, 83*(8), 584–594.
Smith, G. A. (2007). Place-based education: Breaking through the constraining regularities of public school. *Environmental Education Research, 13*(2), 189–207.
Smith, G. A. (2016). Michigan's Great Lakes stewardship initiative: A statewide place-based education effort to connect schools to communities. *LEARNing Landscapes, 10*(1), 289–306.
Smith, G. A., & Sobel, D. (2010). *Place and community-based education in schools: Sociocultural, political, and historical studies in education*. Routledge.
Sobel, D. (2004). Place-based education: Connecting classroom and community. *Nature and Listening, 4*(1), 1–7.
Sobel, D. (2013). *Place-based education: Connecting classroom and communities* (2nd ed.). Orion.
Stevenson, K. T., Peterson, M. N., Carrier, S. J., Strnad, R. L., Bondell, H. D., Kirby-Hathaway, T., & Moore, S. E. (2014). Role of significant life experiences in building environmental knowledge and behavior among middle school students. *The Journal of Environmental Education, 45*(3), 163–177.
Tytler, R., Symington, D., & Clark, J. C. (2017). Community-school collaborations in science: Towards improved outcomes through better understanding of boundary issues. *International Journal of Science and Mathematics Education, 15*(4), 643–661.

Uztemur, S., & Dere, I. (2022). 'I was not aware that I did not know': Developing a sense of place with place-based education. *Innovation: The European Journal of Social Science Research*, 1–17.

Waite, S. (2013). 'Knowing your place in the world': How place and culture support and obstruct educational aims. *Cambridge Journal of Education, 43*(4), 413–433.

Yemini, M., Engel, L., & Ben Simon, A. (2023). Place-based education – A systematic review of literature. *Educational Review*, 1–21.

Yuzuak, A. V., & Zihni, Z. A. (2022). Students' opinions towards place-based education approach in science education: A case study. *MOJES: Malaysian Online Journal of Educational Sciences, 10*(1), 52–64.

Zeidler, D. (2015). Socioscientific issues. In R. Gunstone (Ed.), *Encyclopedia of science education*. Springer.

Zeidler, D. L., & Kahn, S. (2014). *It is debatable: Using socioscientific issues to develop scientific literacy, K-12*. NSTA Press.

Zeidler, D. L., Sadler, T. D., Simmons, M. L., & Howes, E. V. (2005). Beyond STS: A research-based framework for socioscientific issues education. *Science Education, 89*(3), 357–377.

Chapter 2
Developing Preservice Science Teachers' Sense of Place Through Place-Based Socioscientific Issues Instruction

Nilay Ozturk and Büşra Tuncay-Yüksel

2.1 Sense of Place

The massive acceleration in scientific and technological advancements in the twenty-first century has resulted in some contentious and (sometimes) unforeseen consequences. The damage given to nature by the fossil fuel-based energy production industry, for example, is considered one of the major causes of anthropogenic climate change with some global and local impacts (Friedlingstein et al., 2020; Intergovernmental Panel on Climate Change, 2022). Such contentious and unforeseen consequences of science and technology create the need to educate the next generation with a developed set of skills to protect the environment responsibly and make informed and scientifically sound societal decisions.

Research on place-based education argues that when students engage in direct learning experiences in meaningful places, it creates a sense of attachment to the place, motivating them to take a more active and contributing role in addressing their communities' issues and prompting them to take action (e.g., Avriel-Avni et al., 2010; Morgan, 2011; Powers, 2004; Uzzell et al., 2002). In other words, a sense of place is crucial to fostering environmentally conscious and responsible future generations (Morgan, 2011; Moseley et al., 2015).

N. Ozturk (✉)
Department of Primary Education, Faculty of Educational Sciences & BAUSTEM Center, Bahcesehir University, Istanbul, Türkiye
e-mail: nilay.ozturk@bau.edu.tr

B. Tuncay-Yüksel
Department of Mathematics and Science Education, Faculty of Education, Giresun University, Giresun, Türkiye
e-mail: busra.tuncay@giresun.edu.tr

B. Namdar, E. Karahan (eds.), *Socioscientific Issues Focused Teacher Education*, Contemporary Trends and Issues in Science Education 60,
https://doi.org/10.1007/978-3-031-55233-5_2

Sense of place has been defined by different scholars within different frameworks. Gruenewald (2003) proposed a multidisciplinary framework for place-conscious (sometimes used interchangeably with place-based) education. Gruenewald's framework proposed five dimensions of place for the development of socio-ecological and place-conscious education: (a) the perceptual, (b) the sociological, (c) the ideological, (d) the political, and (e) the ecological. Ardoin (2006) proposed four dimensions of sense of place: (a) biophysical setting (providing context for all interactions), (b) psychological (individual) elements, (c) sociocultural elements, and (d) political-economic elements (see Fig. 2.1). According to Ardoin (2006):

> Sense of place describes the complex cognitive, affective, and evaluative relationships people develop with social and ecological communities through a variety of mechanisms. While these relationships are often believed to mature over an extended period within a specific environmental context, they can also occur in a shorter time period through an intense experience or through a strong functional dependence on a certain type of place. Alternatively, a sense of place can also refer to an array of emotional relationships that enhance connections with a variety of social and ecological places. (pp. 118–119)

Based on the four dimensions proposed by Ardoin (2006), Moseley et al. (2015) developed a place consciousness framework. In that framework, similar to Ardoin (2006), sense of place was described as a multidimensional and complex construct based on not only one element (biophysical) but also three interconnected elements of biophysical, psychological, and sociocultural (Moseley et al., 2015) (see Fig. 2.1). In Moseley's framework, the role of psychological elements in place consciousness is central to an individual's construction of a sense of place. Psychological elements involve safety, comfort, security, peace, love, joy, and happiness, which might be influential in developing a psychological connection to a place, a person, or a group (Moseley et al., 2015). The biophysical, sociocultural, and political dimensions overlap as they pass through these psychological elements, contributing to and strengthening individuals' sense of place consciousness.

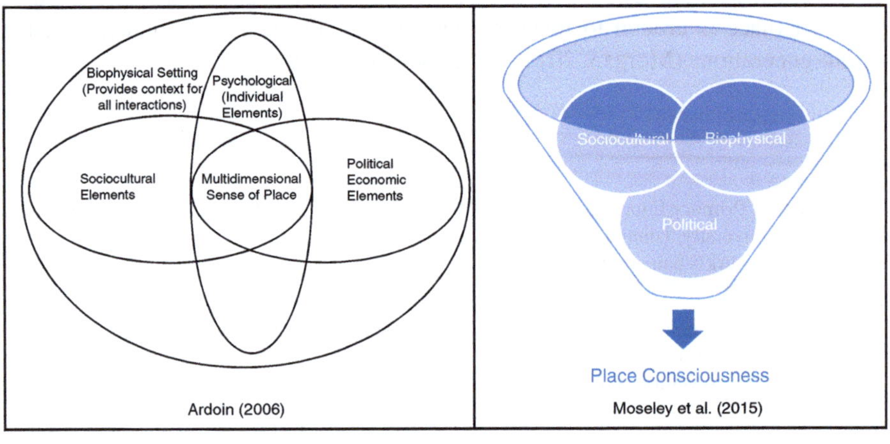

Fig. 2.1 Dimensions of sense of place

Biophysical elements in Moseley's framework include both natural and man-made (built or designed) physical environments. Sociocultural elements are personal experiences and belief systems of an individual that are shaped by cultural influences and commonly shared within families, communities, and societies (Ardoin, 2006; Moseley et al., 2015). Moseley et al. (2015) exemplified the sociocultural elements as the recognition of rituals, traditions, and social structures that connect individuals to a larger community context. Political elements involve issues of social justice and community action. According to Moseley et al. (2015), political elements play an important role in behavior change and commitment to preserving natural resources, supporting the development of a strong sense of place consciousness.

Moseley et al. (2015) argued that the diverse nature of place (including more than the physical space) might make it hard to assume that individuals' sense of place could translate into the conservation and sustainability of natural resources. According to them, in order to accomplish this, it is important to consider the cultivation of a sense of consciousness within the framework of sustainability education, which offers a more comprehensive approach by embracing various aspects (Moseley et al., 2015).

Despite the lack of a consensus on a theoretical framework for place-based education studies, the commonly proposed subdimensions of sense of place are place attachment, place dependence, and place identity (Jorgensen & Stedman, 2001). Place attachment refers to the positive emotional bond that individuals form with a specific location based on their direct experiences (Williams & Vaske, 2003). Place dependence is the perceived strength of a connection to a place, which influences individuals' specific actions (e.g., their preferred activities, such as sports activities in a community park) (Farnum et al., 2005). Place identity refers to the extent to which individuals assimilate themselves with the values and cultures of a particular place (Farnum et al., 2005; Uzzell et al., 2002).

Research studies differ in terms of how they categorize the three subdimensions of sense of place. For example, Jorgensen and Stedman (2001) argue that the three subdimensions explain the meaning of the sense of place well and differentiate place dependence from place attachment. They argue that since place dependence could be negative depending on how individuals perceive the place to be effective for achieving their actions, it is associated with individuals' behaviors; on the other hand, the main focus of place attachment is the emotional aspect experienced by the individuals. Accordingly, Jorgensen and Stedman (2001) investigated individuals' sense of place regarding their place identity, place dependence, and place attachment. Moreover, while some of the researchers suggest that a sense of place is comprised of the two main concepts of place attachment and place meaning (Farnum et al., 2005; Semken & Butler-Freeman, 2008), others consider place dependence and place identity two concepts that form the place attachment (Williams & Vaske, 2003). Despite this variation, it is commonly agreed that sense of place is a multidimensional and complex concept that describes individuals' cognitive and emotional connections to various types of places (e.g., lake, city, neighborhood, etc.) (Hidalgo & Hermandez, 2001) of various sizes (e.g., from a very specific spot to a country) (Vanclay, 2008).

2.2 Socioscientific Issue Context: Dogankent Hydroelectric Power Plant in Giresun

The general SSI context selected in this study is hydroelectric power plants (HPPs) (a particular case of Dogankent HPP located in a small district of Giresun, in this study), which have been discussed as a controversial issue in Türkiye for a long time. In Türkiye, 31% of the total energy production is obtained from HPPs (TEIAS, 2021), and around one-third of the whole energy production comes from the HPPs located in the northern part of the country (Kankal & Akçay, 2019). Giresun, the city where this study took place, is located in the northern part of Türkiye and makes the highest contribution to electricity production through HPPs in that region (Kankal & Akçay, 2019).

The issue of HPPs has generated divergent perspectives and conflicts among different stakeholders, such as environmental groups, power plant companies, farmers, and the local people. On the one hand, HPPs are considered a renewable and clean energy source, as they utilize the power of flowing water to generate electricity. They can contribute to reducing greenhouse gas emissions and dependence on fossil fuels, thus mitigating the impact of climate change (Bayazıt, 2021). Hydroelectric projects can also provide local job opportunities and stimulate economic development in the region. On the other side, there are environmental and social impacts associated with HPPs. Construction of dams and reservoirs can lead to the displacement of communities, loss of agricultural land, and changes in river ecosystems. These projects may also disrupt natural habitats and affect fish populations and other aquatic species. Furthermore, altering the flow of rivers can have downstream impacts on water availability, affecting both human and ecological needs (Kum, 2016).

When examined specifically in the context of the Dogankent HPP, similar effects can be observed. Giresun is a small city with limited economic resources. HPPs like the one in Dogankent create job opportunities for local people. Also, it contributes to the economy of the country by producing electrical energy. On the other hand, (as also stated by the local people during the Dogankent HPP Field Trip; see the following sections of the chapter) the population of *simsir* tree (a plant from the Buxaceae family) has greatly declined, and due to the increased humidity beekeeping has decreased notably in Giresun-Dogankent. Another negative impact of humidity is that agricultural products (for instance, nut cultivation, which is quite important for the region's economy) get moldy while waiting in storage. Moreover, the decreased number of fish population, very hot summer days due to the increased humidity, high voltage power lines passing over and very close to the buildings, and the risk of residential areas being submerged underwater are the other negative impacts of the power plant. So, Dogankent HPP, like other HPPs all around the world, has many ecological, social, ethical, and economic implications.

2.3 Implementation Procedure

2.3.1 Focus and Research Questions

Developing future generations' sense of place has been considered one of the major goals of environmental education (including education for sustainability) practices and curricula to raise conscious, scientifically literate, and equipped individuals to combat the challenges of the twenty-first century (Kudryavtsev et al., 2012). To achieve this goal, supporting in-service and prospective teachers who are the practitioners of environmental education curricular programs is vital. For the effective implementation of curricular programs aiming to improve students' knowledge and skills for pro-environmental behavior, teacher educators need to create opportunities to facilitate teachers' knowledge, skills, attitudes, and beliefs for sustaining the environment, such as developing their sense of place (Moseley et al., 2015; Sarkar & Frazier, 2010).

In this study, we aim to develop a group of preservice science teachers' (PSTs) sense of place through a semester-long undergraduate elective course designed in the context of a locally relevant SSI, namely HPPs. SSIs are science and technology-related social problems without clear-cut solutions such as vaccination and gene therapy (Sadler & Zeidler, 2005). Contextualizing education in these complex and contentious issues has the potential to develop learners' reasoning and argumentation skills, perspective-taking, informed decision-making skills, and eventually functional scientific literacy (Ozturk & Yilmaz-Tuzun, 2017; Sadler et al., 2017). Moreover, SSI-based instruction creates an environment in which concepts and practices are taught through engaging multiple perspectives, not only scientific aspects but also social, economic, ethical, moral, and political aspects of an issue (Sadler & Zeidler, 2005).

The reason we designed the course in the context of an SSI is that research studies highlight the importance of developing a sense of place within a social context (Ardoin, 2006; Gruenewald, 2003). Sense of place does not only involve place identity and attachment to a physical place but also social and cultural connections and taking political action as a community to solve a societal problem (Moseley et al., 2015). SSIs, especially locally-relevant SSIs, facilitate discussions and decision-making related to a societal issue and encourage individuals to generate ideas and solutions by taking into various aspects, which may help them to feel as part of a community and take action. For that reason, we believe that contextualizing the course within an SSI would promote PSTs' sense of place in a more holistic way. The research questions addressed in this study are:

1. How do PSTs' sense of place (i.e., place attachment, place identity, place dependence) change after place-based SSIs instruction?
2. To what extent do PSTs incorporate the dimensions of sense of place (sociocultural, biophysical, political, psychological) for describing their sense of place after place-based SSIs instruction?

2.3.2 Study Context and Participants

This study was implemented in the context of an elective course (i.e., Science and Technology Related Issues) offered for PSTs at a university located in Giresun, Türkiye. The course was developed in accordance with the overall framework established by the Council of Higher Education (2018) and was tailored to the study's objectives. The course was a two-hour course and applied in face-to-face classroom sessions in the 2022–2023 fall semester of the academic year.

The study participants were ten PSTs (five male and five female). Nine of them were sophomores, and one was a senior-grade student in the Science Teacher Education Program. None of the participants or their families' places of residence was Giresun. Therefore, it was their first year in the city owing to distance education applied in Turkish universities during the previous 2 years during the COVID-19 pandemic. Nonetheless, one of the participants was an international student from Turkmenistan who had been living in Giresun since he first came to Türkiye 4 years ago.

2.3.3 Details of the Teaching and Learning Processes

Implementation of the course was realized throughout a 16-week time period. The foci of the weeks and the learning and teaching activities realized in the course sessions are summarized in Table 2.1. Detailed information is provided in the following sections.

2.3.3.1 Warm-Up Weeks

The first two weeks of the course were devoted to the introduction of the course and concepts related to science and technology. The third week of the course mainly aimed at helping the PSTs to be able to develop an awareness of the interrelationships among the concepts that are related to the "Science and Technology Related Issues" course. For this, brainstorming and concept mapping techniques were used. More specifically, the PSTs first did brainstorming and wrote all of the related concepts that came into their minds on the classroom board. Then, they created concept maps by using the concepts on the classroom board as well as any other concepts they thought would be related to the course. Finally, discussions were made on the concept maps created by the PSTs. Science education approaches that focus on science-technology-society- [environment] relationships (STS[E]) and their places in Turkish middle school curricula were also explained and discussed in the third week of the course. In this third week, the PSTs were introduced to the concept of SSIs, but no information was given about the meaning of SSIs. In fact, the first three weeks of the course were a kind of "warm-up" period for the PSTs in which they

Table 2.1 Details of the course

Week		Focus	Teaching and learning activities
Warm-up weeks	W1	Introduction	Introduction to the course: Course syllabus & PST expectations
	W2	Concepts related to science and technology	Small group and whole class discussions about science and technology-related terminologies
	W3	Science education approaches that focus on science-technology-society-environment relationships (i.e., STS, STSE, SSIs)	Brainstorming and concept-mapping
Week 4		**Data collection**	
SSI and sense of place weeks	W5	Introduction to SSIs and SSI-based instruction	Class activity on SSIs scenarios about organ donation
	W6	The importance of SSIs in science education Interdisciplinary nature of SSIs	Examination of the teaching modules of a national project on SSIs instruction Class activity on SSIs scenarios *Homework for the next week:* Research on Dogankent HPP
Pre-field trip weeks	W7	Doganken HPP	Class discussions on the homework
	W8	**Exam week** (No class)	
	W9	Preparation for the field trip	Class discussions on the field trip Completion of the pre-field trip worksheets
Field trip week	W10	Field trip to Dogankent HPP and the Dogankent district	Explanations of the chief engineer of the HPP, discussions with the HPP staff, PST observations Interviews with the local people Completion of the field trip worksheets
Post-field trip weeks	W11	Reflections on the field trip	Class discussions about the PSTs' field trip experiences and influences on their feelings and thoughts about the SSI & sense of place towards Giresun
	W12	Documentary named "Citizen Mustafa [Vatandaş Mustafa]" Similarities and differences between the SSI in the documentary and the SSI examined throughout the course	Documentary watching Classroom discussions *Homework for the next week:* Reflections on the documentary and classroom discussions
	W13 & 14	Perceptions about the Dogankent HPP as SSI Reflections on the influences of the course & field trip on the PSTs including their sense of place	Student presentations and classroom discussions
Week 15		**Data collection**	
Week 16		**Exam week** (No class)	

practiced small group/class discussions. These "warm-up weeks" were considered to be important for the study because, due to the Covid-19 pandemic, the PSTs had been away from a face-to-face classroom environment for two academic years. It was clear that they needed time to get to know each other and feel a "safe" classroom environment where they could freely express their ideas and feelings.

2.3.3.2 Data Collection (Pre-test)

After the "warm-up weeks," the Sense of Place Scale (Jorgensen & Stedman, 2001) was applied as a pre-test. The Sense of Place Scale includes twelve self-report items that measure the sense of place in three components (i.e., place identity, place attachment, and place dependence). The scale was originally developed to collect data from lakeshore property owners in Wisconsin, USA. In this study, item statements in the scale were translated into Turkish and adapted to measure the PSTs' sense of place with regard to Giresun. The two authors first translated the items in the scale from English to Turkish individually and afterward came together to discuss whether the translated items sufficiently captured the meaning in Turkish. That was followed by getting an English-language expert opinion on the first translated version of the scale. Then, with the aim of ensuring the validity of the scale, expert opinion was taken from a colleague who had been working on preservice teachers' sense of place for checking the appropriateness of the translation and adaptation of the original scale. That is, the colleague checked the items in terms of their sufficiency in revealing respondents' sense of place regarding Giresun without losing the main focus in the original scale items. Based on the revisions taken, the authors finalized the scale. For the reliability analysis, the authors were cautious about running factor analysis and computing Cronbach's alpha coefficient. The reason is that, for factor and reliability analysis, it is usually recommended to have at least 300 cases or 10 to 1 ratio of cases to items (Tabachnick & Fidell, 2007). Due to the small group of participants (10 PSTs) in the present study, we avoided running factor analysis and computing reliability coefficients.

2.3.3.3 Socioscientific Issues and Sense of Place Weeks

The following two weeks of the course (the fifth and the sixth week) were devoted to SSIs and their use in science education. More specifically, in the fifth week of the course, features of SSIs and differences in SSI approach from the STS[E] approach were discussed, and examples of local, regional, national, and global SSIs were provided. The importance of SSI in science education was also discussed with references to the literature and teaching modules of a national project funded by the Scientific and Technological Research Council of Türkiye. Moreover, the interdisciplinary nature of SSIs and decision-making processes was explained via the SEE-SEP (science, environment, ethics/morality, sociology/culture, economy, and policy) model (Chang Rundgren & Rundgren, 2010). Based on the importance of practice and experience in learning and teaching in the SSI context (Zeidler & Nichols, 2009), the PSTs were provided with two SSIs activities about organ donation (Tuncay-Yüksel & Cansız, 2020; Zeidler et al., 2013) and pre-implantation genetic diagnosis (Cebesoy, 2020) in the fifth and sixth weeks of the course, respectively. These activities provided participants with insight into SSIs, as well as direct experience with reasoning and decision-making about SSIs contexts.

The fifth and sixth weeks of the course were also course sessions in which the PSTs were encouraged to think about the relationships between SSIs and sense of place in the context of the issue of HPPs in Giresun. Rather than giving explicit explanations about those relationships (i.e., relationships between SSIs and sense of place), the instructor put effort into utilizing an implicit approach. Accordingly, smooth transitions between successive topics/issues were ensured throughout the course. For instance, in the fifth week of the course, classroom discussions were made about the PSTs' thoughts and feelings about being in Giresun by asking questions such as "Do you feel that you belong to Giresun and/or the region in which it is located?", "Would you like to live in Giresun or in this region after you graduate?" etc. These questions were anticipated to guide the PSTs to think about their relationships with Giresun and sense of place they perceive regarding the city as well as the factors that may be influential on those perceptions. Then, in the sixth week, homework was given to the PSTs in which they were asked to research and think about the stakeholders of the issue of Dogankent-HPP and the factors that may be influential with regard to the reasoning and decision-making processes about the issue.

2.3.3.4 Pre-field Trip Weeks

A field trip to a local HPP in Giresun was considered an important element of the course in that it was believed to contribute to the PSTs' sense of place in terms of the SSI examined throughout the course. As previously announced in the course syllabus, the field trip was scheduled two weeks after the midterm examination week (week 8). The weeks before the field trip were utilized to prepare the PSTs for the field trip and make bureaucratic arrangements, such as taking official permissions from the relevant rectorate units and organizing the university shuttle service. During all steps of the field trip (including pre-field trip and after-field trip steps), close attention was paid to the issues that should be considered while organizing, conducting, and finalizing a field trip (Eshach, 2007).

In the seventh week of the course, classroom discussions were made about the PSTs' homework in order to deepen the participants' understanding of the various aspects of the HPPs in Giresun and their relationships to their (daily) lives. In the course session after the midterm exam (week 9), the instructor and the PSTs met to discuss the details of the field trip (e.g., the purpose of the field trip, general characteristics of the place (Dogankent District and the HPP in Dogankent) to be visited, questions to be asked to local people living in Dogankent and the staff working at the HPP, etc.). Moreover, the PSTs completed a pre-field trip worksheet in which they responded to five open-ended questions about their personal experiences (if any) with an HPP, expectations from the field trip, ideas about the possible influences of the Dogankent HPP on the district, perceptions about relationships of the HPP issue to their (daily) lives, and feelings about the planned field trip.

2.3.3.5 Field Trip Week

In the field trip week of the course (week 10), the PSTs and the instructor met at the predetermined meeting place at the university campus. As observed in their responses to the questions in the pre-field trip sheet, the PSTs were very enthusiastic and excited about the field trip because none of them had visited an HPP before. The person who was contacted in advance (the chief engineer of the power plant) welcomed and accompanied the group throughout the field trip. He gave information about the HPP (e.g., its history, features, working principle, etc.) and answered the PSTs' questions (questions prepared in advance and questions that spontaneously came into their minds during the trip). These questions were generally about the possible influences of the HPP on the local environment (e.g., effects of the HPP on the river ecosystem and the region in general) and the people living there (e.g., effects on the economy of the people, their health, etc.).

After the visit, the instructor and the PSTs went to the settlement of the Dogankent district in order to examine the issue of HPP from a wider perspective. First, group discussions were held about the answers of the HPP staff to the PSTs' questions, and questions to be asked to the local people living in the district were reviewed. Then, the PSTs had a short tour of the district to have an idea about the district and the aspects that should be considered while reasoning about the issue of HPP (e.g., the distance of the settlement from the power plant, features of the ecosystem around the settlement, etc.). As in many small districts of Türkiye, people were outside. Therefore, it was quite easy to find people (especially men) and ask questions about the HPP. The PSTs first introduced themselves and explained that they wanted to ask questions about the HPP for a course they were taking at the university. Local people were very hospitable. They continuously offered tea to the PSTs, and many of them gave permission to record their answers to the questions.

Following the interviews with the local people, the PSTs and the instructor came together to summarize and critique the day, including their experiences in the HPP campus and the Dogankent district. Upon the PSTs' request, a 20-min free time was given for buying local food. In fact, this free time was a chance for the PSTs to contact the local people in a more "informal" way. On the way to the university campus, the PSTs completed and assigned the worksheets that included questions about the HPP and their experiences during the field trip.

2.3.3.6 Post-field Trip Weeks

The next week of the course (11th week) was devoted to discussions about the field trip, including the PSTs' experiences during the field trip, influences (if any) of the field trip on their feelings and thoughts about the issue, and their sense of place towards Giresun. In the following week (12th week), a documentary named "Citizen Mustafa [Vatandaş Mustafa]" (Kazmaz, 2015) was watched in the classroom. The documentary is about the struggle of people (local people as well as other people from the country and the world) against the construction of an HPP in the Fırtına Basin-northern Türkiye. After watching the documentary, homework was assigned to the PSTs in which they were asked to reflect on the documentary and compare the similarities and differences between the HPP issue in the documentary and Dogankent HPP. Then, the class was informed about the presentations they would make in the following weeks of the course. The presentations were expected to be organized in two main parts: (1) The SSI (HPPs in Giresun in general and the Dogankent HPP in particular) they examined throughout the course (e.g., why it is considered as a SSI, aspects that should be considered while reasoning and decision making about the issue), (2) Reflections about the field trip and the course including the influences (if any) on their perceptions, viewpoints, feelings, etc.

Classroom discussions were made during and after the PSTs' presentations. The 14th week of the course was also a kind of general evaluation of the course where most of the PSTs' expressed their satisfaction with and happiness for taking the course. Many of the PSTs stated that the course was very beneficial both for improving skills necessary for their academic and professional life and their skills in general. Statements of one of the PSTs exemplify this: "This course has had many positive effects on us. First of all, we think that doing activities in each lesson, having discussions, and sharing our ideas about them [the issues discussed in the course] developed our reasoning skills. Going to Dogankent and chatting with people also had positive effects. It developed our ability to communicate and empathize with people." The PST's statements also implied some changes in their sense of place regarding Giresun: "After the field trip, our perspective towards Giresun changed in a positive way. At first, Giresun didn't mean much to us. But after our trip, we realized that we felt a little closer to here. We started to embrace Giresun's problems more."

2.3.3.7 Data Collection (Post-test)

In the last week of the course (15th week), the Sense of Place Scale (Jorgensen & Stedman, 2001) was applied as a post-test. In addition, the PSTs were asked four open-ended questions related to their sense of place towards Giresun. The four open-ended questions were: (1) Can you establish a connection between your identity and the city of Giresun? How much does the city of Giresun reflect your identity? (2) Do you feel relaxed and happy when you are in Giresun? If yes, why? If not, why? (3) Are you satisfied with being in Giresun? (4) Do you feel a sense of belonging to Giresun and/or this region?

2.4 Outcomes

2.4.1 *Preservice Science Teachers' Sense of Place After the Implementation*

According to the descriptive data obtained through the Sense of Place Scale (Jorgensen & Stedman, 2001), the PSTs scored highest on the place dependence dimension (pre: M = 3.25, SD = .28; post: M = 3.18, SD = .40) both before and after the implementation. Mean scores associated with place identity (pre: M = 2.62, SD = .65; post: M = 3.00, SD = .64) and place attachment (pre: M = 2.59, SD = 1.29; post: M = 2.85, SD = .94) were above 2.5 on the 1–5 scale both before and after the implementation. The overall sense of place mean scores before and after the implementation were 2.82 (SD = .62) and 3.01(SD = .56), respectively.

The Wilcoxon Signed Rank Test was conducted to test whether there was a significant change between pre and post-measurements of place dependence, place identity, and place attachment. The results revealed a statistically significant increase with a large effect size for the place identity dimension (z = 2.69, p = .007, r = 0.60). There was no statistically significant difference between the pre-post measurements of place dependence and place attachment (Table 2.2). According to these results, PSTs' sense of place identity statistically significantly developed after the place-based SSIs-based instruction; however, the increase in PSTs' sense of place attachment was not statistically significant. Moreover, there was a decrease in PSTs' sense of place dependence, but this decrease was also not statistically significant.

2.4.2 *Preservice Science Teachers' Incorporation of Sense of Place Dimensions After the Implementation*

Answers to the four open-ended questions were coded to explore which dimensions of sense of place (i.e., biophysical, sociocultural, political, and psychological) were included in PSTs' descriptions of their negative or positive sense of place towards Giresun (Moseley et al., 2015) (Table 2.3). Coding of the PSTs' answers was first carried out independently by the two researchers. Comparisons between their coding revealed 95% inter-rater consistency. After discussions, all discrepancies were resolved, and a 100% agreement was reached. Results showed that while describing

Table 2.2 Pre to post changes in the PSTs' sense of place

Sense of place dimension	*z*	*P*	Effect size (Cohen, 1988)
Place identity	2.69	.007*	0.60
Place attachment	1.06	.285	
Place dependence	0.34	.732	

*Significant at 0.05 level

Table 2.3 Inclusion of sense of place dimensions in PST answers

Dimension/Code	Code description	Frequency
Biophysical		
Natural beauty	Physical environment including sea, forests, creeks	8
Climate	Climate patterns in Giresun (rainy, cloudy, gloomy, etc.)	2
Opportunities	Lack of (places for social activity and educational) opportunities due to being a small city	5
Sociocultural		
Having friends	Having friend groups living in Giresun	3
Missing hometown	Having a hard time feeling connected to Giresun due to missing family, family traditions, etc.	2
Lifestyle/worldview	Incompatibility between the life in Giresun and PSTs' lifestyle and worldview	2
Student identity	Sense of identity towards Giresun only because PST is going to college in that city	1
Connection to a larger community	Feeling a part of a larger community living in Giresun	4
Political		
Community action	Acting all together to solve a problem in Giresun	1
Psychological		
Unhappy	Feeling unhappy to be in Giresun	1
Bored	Feeling bored to be in Giresun due to the lack of social activities	3
Peaceful	Feeling peaceful because Giresun is not a big city	2
Comforted	Feeling comforted because Giresun is not a chaotic city	4
Love	Love of Giresun	1
Familiarity	Feeling familiar because of the already time spent in Giresun	3
Freedom	Feeling free to be able to live on their own	1

Note: 10 out of 10 PSTs responded to the open-ended questions after implementation

their sense of place towards Giresun, PSTs used biophysical dimensions the most. Namely, they stated that the natural beauty of Giresun (i.e., the sea, creeks, forests, trees, parks, etc.) is a reason for them to feel connected to Giresun. However, some of the PSTs mentioned that they do not like the gloomy weather and the very limited social activity opportunities in Giresun. The following two quotes represent natural beauty and climate-related PST responses:

> I am happy to be in Giresun. I feel connected to it through its nature, creeks, prairie, and mountains. I will miss this place when I am back home. (PST3)
>
> Giresun does not reflect my identity because Giresun is very gloomy, very cold, and I am not that kind of person. (PST5)

Moreover, PSTs incorporated sociocultural dimensions with a similar frequency of biophysical dimension. They referred to having friends and feeling a part of a larger community in Giresun. On the other hand, as mentioned in PSTs' answers, since all of the PSTs are living in Giresun for college, Giresun is not their hometown, and due to that, it is not easy for them to feel fully connected to Giresun. They mentioned that they miss their families and the time they have with their families in their hometown. In addition, one of the PSTs expressed that his lifestyle is different from the local people living in Giresun, so he does not like to be in Giresun. One of the PSTs stated:

> I am not happy being in Giresun because there is quite a difference between the life I used to have and the life I have here. There are cultural and mindset differences between where I come from and here. I consider cities, where people are more free, and nobody interferes with others' lives, closer to myself. (PST7)

The psychological dimension of a high or low sense of place articulated by the PSTs varies from unhappy and bored to peaceful and comforted. With the highest frequency, PSTs mentioned that they feel comforted because Giresun is a small and sparsely populated city with no heavy traffic. For example, one of the PSTs stated that: "I am glad that I lead a comfortable life here because Giresun is a small city" (PST8). However, according to some PSTs, Giresun is a very boring place where they do not have space to entertain and socialize.

The political dimension was the least to be included in PST answers. Only one PST referred to taking community action to solve environmental problems in Giresun. She brought up the pollution on the seashore and mentioned that she is trying to do her best to clean the garbage on the beach. She also mentioned that there needs to be stronger action that involves everyone living in the city to find a durable solution to the garbage problem on the beach. She articulated that:

> I feel sad when I see excessive garbage on the beach. I do my best for a clean environment, but it's not enough with just my efforts, we need to act together. That's why Giresun reflects not only my identity but also the identity of everyone living in it as a whole.

The data showed that there was a significant increase in PSTs' sense of identity after implementation. However, no significant changes were obtained for a sense of dependence and sense of attachment. One of the reasons could be that Giresun was not the hometown of any of the participating PSTs. They currently reside in Giresun because they attended university there, and it is uncertain whether they will continue to live there after graduation. Also, the PSTs are sophomores, and in the first three semesters of their undergraduate program, the courses were online due to COVID-19 pandemic restrictions. So, the semester that this study was implemented was the first semester they took in-person classes in Giresun. For these reasons, the course in this study could have failed to develop PSTs' sense of dependence and sense of attachment towards Giresun. Outcomes obtained for the second research question revealed that PSTs have different motives and incorporate different dimensions of sense of place while describing their sense of place towards Giresun. Biophysical and psychological dimensions are the ones that are mostly incorporated in PST responses. However, the political dimension was the least influential dimension in shaping PSTs' sense of place towards Giresun. This might be an implication for teacher educators that teacher education courses aiming to develop a sense of place in the context of a controversial SSI might involve more emphasis on political dimensions, including issues of justice and community action.

2.5 Transfer to the Global Context

Energy production is one of the vital needs for every country. HPPs are widely used to produce energy and (as mentioned in detail above) create controversial discussions due to their impacts all around the world. Electricity generated through HPPs in Türkiye constitutes around 20% of the total production in 2022, while electricity generated through coal and fuel oil is higher in percentage (Turkish Ministry of Energy and Natural Resources, 2023). There are countries in different regions of the world that obtain a higher proportion of their electricity from HPPs compared to Türkiye. For example, in countries such as China, Brazil, and the USA, electricity production is much more dependent on HPPs (World Energy Council Turkey, 2021). Moreover, in recent years, there has been an increasing number of ongoing HPP construction projects, particularly in underdeveloped regions like South America and Africa, which sometimes create conflict situations (Zarfl et al., 2015). Therefore, the issue of HPPs is familiar to several other regions in the world and thus could be used as an SSI context with multiple dimensions for developing preservice and in-service teachers' sense of place.

The teaching approach used in the semester-long undergraduate course could also be transferable to other contexts worldwide. The course designed for this study could be offered as an elective or must-course in teacher education programs in different countries. Alternatively, the implementation (i.e., field trip) within this course can be incorporated into courses with themes such as scientific literacy, argumentation/reasoning, SSI, or as part of a science methods teaching course. As implemented in this study, teacher educators might design their course around a local HPP and its impacts on different stakeholders and groups. We recommend organizing field trips (multiple times if possible) to provide opportunities for preservice teachers to engage with the local people, the power plant workers (engineers and scientists), local environmental groups and NGOs, and policymakers. As mentioned in detail in the above sections, the course should be designed to involve pre-trip, during-trip, and post-trip activities. Before the field trip, it is important to provide preservice teachers with the opportunity to engage in prior research and stimulate their thinking process. After the field trip, the course should let various discussions where preservice teachers reflect, share their positions, and help them feel as part of the problem and act for the resolution. Activities planned for the field trip provide a significant opportunity for preservice teachers to understand the local problem from various perspectives (e.g., social, political, economic, etc.). These activities should encourage preservice teachers to make connections with the local people, such as conducting interviews with people living nearby the power plant. In addition, they should be able to ask technical questions of the staff at the power plant, such as engineers and scientists. Depending on the sociocultural context where the course is implemented, teacher educators might focus on the impacts (and related aspects) of the power plant. For example, if the power plant is constructed in an environmentally sensitive area, the course might highlight the environmental aspect. Likewise, if the HPP is the only income source for people living in a town, then the implementation might push preservice teachers to answer economy-focused discussion questions. We believe that incorporating controversial issues such as HPPs in teacher education courses has a major potential to make preservice teachers feel connected to their local area, which eventually promotes them to protect their environment and act to find a solution to local problems with the larger community.

2.6 Conclusion

This study aimed to develop PSTs' sense of place during a semester-long SSIs-based instruction in a city located in the northern part of Türkiye. The issue of HPPs was selected as the SSI context. A 16-week-long elective course was designed consisting of classes focusing on the theoretical background of SSIs and SSIs-based teaching, discussions about varying SSIs relevant to the PSTs, a detailed examination of the issue of HPPs, a field trip to a local HPP, and opportunities for PSTs to reflect on and share their experiences, feelings, ideas, and so on. According to pre-post measures of PSTs' sense of identity, sense of dependence, and sense of attachment, there was a statistically significant increase in PSTs' sense of place, but the findings did not show any significant increase in the sense of dependence and sense of attachment. Moreover, after the implementation, analysis of PSTs' responses to a sense of place questions showed that the PSTs mostly incorporated biophysical and psychological dimensions but almost no political dimension while describing their sense of place towards Giresun. We believe that contextualizing the course in a local HPP issue helped PSTs to understand the impacts of a local SSI and foster them to feel responsible for finding a solution as a part of the larger community living in the region. We believe that the field trip and the whole course design could be insightful for other teacher educators in different regions of the world and transferable to several other contexts.

Acknowledgments We would like to thank the preservice science teachers for their cooperation and contributions to the course and the study.

References

Ardoin, N. (2006). Toward an interdisciplinary understanding of place: Lessons for environmental education. *Canadian Journal of Environmental Education, 11*(1), 112–126.

Avriel-Avni, N., Spektor-Levy, O., Zion, M., & Levi, N. (2010). Children's sense of place in desert towns: A phenomenographic enquiry. *International Research in Geographical and Environmental Education, 19*(3), 241–259.

Bayazıt, Y. (2021). The effect of hydroelectric power plants on the carbon emission: An example of Gokcekaya Dam, Turkey. *Renewable Energy, 170*, 181–187.

Cebesoy, Ü. B. (2020). SEE-SEP model in teaching socioscientific issues [Sosyobilimsel konuların öğretiminde SEE-SEP modeli]. In M. Genç (Ed.), *Socioscientific issues from theory and to practice* [Kuramdan uygulamaya sosyobilimsel konular] (pp. 155–180). Nobel Academic Yayıncılık.

Chang Rundgren, S. N., & Rundgren, C. J. (2010). SEE-SEP: From a separate to a holistic view of socioscientific issues. *Asia-Pacific Forum on Science Learning and Teaching, 11*(1), Article 2.

Cohen, J. W. (1988). *Statistical power analysis for the behavioral sciences* (2nd ed.). Lawrence Erlbaum Associates.

Council of Higher Education. (2018). *Fen Bilgisi Öğretmenliği Lisans Programı* [Undergraduate Program for Science Teacher Education]. https://www.yok.gov.tr/Documents/Kurumsal/egitim_ogretim_dairesi/Yeni-Ogretmen-Yetistirme-Lisans-Programlari/Fen_Bilgisi_Ogretmenligi_Lisans_Programi.pdf

Eshach, H. (2007). Bridging in-school and out-of-school learning: Formal, non-formal, and informal education. *Journal of Science Education and Technology, 16*(2), 171–190.

Farnum, J., Hall, T., & Kruger, L. E. (2005). *Sense of place in natural resource recreation and tourism: An evaluation and assessment of research findings*. General technical report PNW-GTR-660: US Department of Agriculture, Forest Service, Pacific Northwest Research Station.

Friedlingstein, P., O'Sullivan, M., Jones, M. W., Andrew, R. M., Hauck, J., Olsen, A., Peters, G. P., Peters, W., Pongratz, J., Sitch, S., Quéré, C. L., Canadell, J. G., Ciais, P., Jackson, R. B., Alin, S., Aragão, L., Arneth, A., Arora, V., Bates, N. R., et al. (2020). Global carbon budget 2020. *Earth System Science Data, 12*(4), 3269–3340.

Gruenewald, D. A. (2003). Foundations of place: A multidisciplinary framework for place-conscious education. *American Educational Research Journal, 40*(3), 619–654.

Hidalgo, M. C., & Hernández, B. (2001). Place attachment: Conceptual and empirical questions. *Journal of Environmental Psychology, 21*(3), 273–281.

Intergovernmental Panel on Climate Change (IPCC). (2022). *Climate Change 2022: Impacts, adaptation and vulnerability report. Contribution of working groups II to the sixth assessment report of the intergovernmental panel on climate change*. Cambridge University Press.

Jorgensen, B. S., & Stedman, R. C. (2001). Sense of place as an attitude: Lakeshore owners attitudes toward their properties. *Journal of Environmental Psychology, 21*(3), 233–248.

Kankal, M., & Akçay, F. (2019). Doğu karadeniz havzası hidroelektrik enerji durumunun incelenmesi. *Niğde Ömer Halisdemir Üniversitesi Mühendislik Bilimleri Dergisi, 8*(2), 892–901.

Kazmaz, R. (2015). *Vatandaş Mustafa (Documentary)* [Video]. YouTube. https://www.youtube.com/watch?v=4Rqa2eeUadQ

Kudryavtsev, A., Stedman, R. C., & Krasny, M. E. (2012). Sense of place in environmental education. *Environmental Education Research, 18*(2), 229–250.

Kum, G. (2016). The influence of dams on surrounding climate: The case of Keban Dam. *Gaziantep University Journal of Social Sciences, 15*, 193–204.

Morgan, A. (2011). Place-based education versus geography education? In G. Butt (Ed.), *Geography, education and the future* (pp. 84–108). Continuum International Publishing Group.

Moseley, C., Desjean-Perrotta, B., & Kharod, D. (2015). Sense of place: Is it more than a connection to a physical place? In S. K. Stratton, R. Hagevik, A. Feldman, & M. Bloom (Eds.), *Educating science teachers for sustainability* (pp. 31–48). Springer.

Ozturk, N., & Yilmaz-Tuzun, O. (2017). Preservice science teachers' epistemological beliefs and informal reasoning regarding socioscientific issues. *Research in Science Education, 47(6)*, 1275–1304.

Powers, A. (2004). Teacher preparation for environmental education: Faculty perspectives on the infusion of environmental education into preservice methods courses. *Journal of Environmental Education, 35*(3), 3–10.

Sadler, T. D., & Zeidler, D. L. (2005). Patterns of informal reasoning in the context of socioscientific decision making. *Journal of Research in Science Teaching, 42*(1), 112–138.

Sadler, T. D., Foulk, J. A., & Friedrichsen, P. J. (2017). Evolution of a model for socio-scientific issue teaching and learning. *International Journal of Education in Mathematics, Science and Technology, 5*(2), 75–87.

Sarkar, S., & Frazier, R. (2010). Place-based inquiry: Advancing environmental education in science teacher preparation. In A. M. Bodzin, B. S. Klein, & S. Weaver (Eds.), *The inclusion of environmental education in science teacher education* (pp. 159–171). Springer.

Semken, S., & Butler-Freeman, C. (2008). Sense of place in the practice and assessment of place-based science teaching. *Science Education, 92*, 1042–1057.

Tabachnick, B. G., & Fidell, L. S. (2007). *Using multivariate statistics*. Pearson.

Tuncay-Yüksel, B., & Cansız, N. (2020). Socioscientific reasoning [Sosyobilimsel konularda muhakeme]. In M. Genç (Ed.), *Socioscientific issues from theory and to practice* [Kuramdan uygulamaya sosyobilimsel konular] (pp. 181–209). Nobel Akademik Yayıncılık.

Turkish Electricity Transmission Corporation (TEIAS). (2021). *Türkiye elektrik iletim-uretim 2021 yili istatistikleri.* https://www.teias.gov.tr/Türkiye-elektrik-uretim-iletim-istatistikleri

Turkish Ministry of Energy and Natural Resources. (2023). *Electricity*. https://enerji.gov.tr/infobank-energy-electricity

Uzzell, D., Poi, E., & Badenas, D. (2002). Place identification, social cohesion, and environmental sustainability. *Environment and Behavior, 34*(26), 26–53.

Vanclay, F. (2008). Place matters. In F. Vanclay, M. Higgins, & A. Blackshaw (Eds.), *Making sense of place. Exploring concepts and expressions of place through different senses and lenses* (pp. 3–11). National Museum of Australia Press.

Williams, D. R., & Vaske, J. J. (2003). The measurement of place attachment: Validity and generalizability of a psychometric approach. *Forest Science, 49*, 830–840.

World Energy Council Turkey. (2021). *2021 hydroelectric situation report summary*. https://www.dunyaenerji.org.tr/2021-hidroelektrik-durum-raporu-ozeti/#:~:text=%C3%87in%2C%20370%20GW'%C4%B1n%20%C3%BCzerinde,ilk%20be%C5%9Fin%20geri%20kalan%C4%B1n%C4%B1%20olu%C5%9Fturuyor

Zarfl, C., Lumsdon, A. E., Berlekamp, J., Tydecks, L., & Tockner, K. (2015). A global boom in hydropower dam construction. *Aquatic Sciences, 77*, 161–170.

Zeidler, D. L., & Nichols, B. H. (2009). Socioscientific issues: Theory and practice. *Journal of Elementary Science Education, 21*, 49–58.

Zeidler, D. L., Herman, B. C., Ruzek, M., Linder, A., & Lin, S. S. (2013). Cross-cultural epistemological orientations to socioscientific issues. *Journal of Research in Science Teaching, 50*(3), 251–283.

Chapter 3
Preservice Science Teachers' Socioscientific Orientations Through Placed-Based Socioscientific Issues Teaching: Air Pollution in Ankara

Süleyman Çite and Selçuk Şahingöz

3.1 Introduction

Promoting active and responsible citizenship to foster a sustainable and inclusive global economy and society is a key focus of national science education standards (MEB, 2018; NGSS, 2013) and international non-profit organizations (e.g., OECD, 2018). The development of such citizens is best nurtured through immersing learners in societal problems, as resolving these issues requires critical thinking and consideration of social, ethical, and moral dimensions (Bencze et al., 2020; Herman et al., 2021).

Efforts in science education worldwide emphasize the achievement of functional scientific literacy for all (Hodson, 2009; OECD, 2017; Siarova et al., 2019). Functional scientific literacy equips learners with the ability to effectively engage with complex societal issues rooted in science, requiring consideration of social, economic, ethical, and cultural aspects (Fortus et al., 2022; Sadler, 2004).

While anchoring instruction around societal issues and providing diverse perspectives have been linked to improved academic achievement, this alone is insufficient for cultivating the competencies necessary for functional scientific literacy (Zangori et al., 2020; Zeidler et al., 2019). Science educators argue for the inclusion

S. Çite (✉)
Department of Mathematics and Science Education, Faculty of Education,
Kastamonu University, Kastamonu, Türkiye
e-mail: suleymancite@kastamonu.edu.tr

S. Şahingöz
Department of Mathematics and Science Education, Faculty of Education,
Gazi University, Ankara, Türkiye
e-mail: selcuksahingoz@gazi.edu.tr

B. Namdar, E. Karahan (eds.), *Socioscientific Issues Focused Teacher Education*, Contemporary Trends and Issues in Science Education 60,
https://doi.org/10.1007/978-3-031-55233-5_3

of place-based teaching approaches within SSI instruction to foster these necessary competencies (Herman et al., 2021; Powell et al., 2021). By engaging with individuals directly impacted by the issue or those influenced by it, learners can effectively develop the affective characteristics required for informed decision-making and taking positions on the issue. Thus, the purpose of this chapter is to share the experiences of preservice science teachers as they confront the issue of air pollution in Ankara. Specifically, we aim to understand and document the experiences of preservice science teachers through their engagement in various place-based activities related to air pollution in Ankara, focusing on their socioscientific orientations.

3.1.1 Socioscientific Orientation as a Mean of Documenting Changes

Examining the outcomes of learners' engagement in SSI-based instruction has been a focal point for science education researchers (Zeidler et al., 2019). Previous research on SSI-based instruction has primarily focused on learning outcomes such as content knowledge development, improved understanding of the nature of science, and enhanced reflective thinking (Sadler, 2004). While these outcomes are valuable, scholars argue that SSI-based instruction should also foster affective characteristics necessary for active and responsible citizenship (Fortus et al., 2022; Sadler et al., 2007).

To address this challenge, Sadler et al. (2007) introduced the concept of socioscientific reasoning (SSR), which encompasses complexity, perspective-taking, inquiry, skepticism, and understanding the affordances and limitations of science. While SSR captures substantial learning gains related to SSI engagement, research suggests that detecting these gains requires deep and extended SSI learning experiences (Kinslow et al., 2019; Romine et al., 2017).

Herman et al. (2021) proposed a construct called socioscientific orientations to capture changes in undergraduate students' perspectives resulting from a six-week-long, place-based SSI instruction. Socioscientific orientations encompass ecological worldviews (interconnectedness and sustainable development), social and moral compassion (moral and ethical sensitivity, perspective taking, and empathetic concerns), socioscientific accountability (feeling of responsibility and willingness to act), and scientific evidence views (considering the affordances and constraints of scientific evidence). The socioscientific orientations approach is more suitable for our study, as our SSI instruction centers around the environmental issue of air pollution in Ankara, incorporating place-based elements. Additionally, compared to the SSR framework, the socioscientific orientation approach allows for a more detailed analysis of changes in learners' perspectives, enabling us to capture subtle shifts that may be overlooked when focusing solely on SSR components.

In addressing environmental challenges, it is essential for students to understand that sustainable solutions require the removal of anthropocentrism from the environment—an aspect covered by the interconnectedness and sustainable development dimension of the socioscientific orientations framework. In our study, we aim to explore our participants' perspectives on how they approach the resolution of the air pollution issue in Ankara. Specifically, we are interested in understanding to what extent learners' proposed solutions consider the reciprocal relationship between humans and nature.

Ankara's air pollution issue entails numerous social and ethical considerations that must be addressed. Learners should approach the problem through the perspectives of those affected by it and those directly contributing to its emergence. Furthermore, different socioeconomic groups in Ankara experience varying impacts from air pollution, necessitating the consideration of social and economic injustices that may arise from proposed solutions. Empathy is a crucial element to be cultivated.

We also seek to examine whether students feel accountable for the emergence and resolution of Ankara's air pollution problem. A sense of responsibility increases the likelihood of dedicating time and effort to addressing the issue. Therefore, we aim to understand if learners perceive any responsibility for resolving the air pollution problem in Ankara and whether they are willing to take action towards its solution due to engaging in place-based SSI instruction.

Any SSI issue can only be effectively resolved by considering the affordances and limitations of scientific evidence. Scientific data undoubtedly plays a significant role in addressing Ankara's air pollution problem, but implementation must also consider the city's social, cultural, and economic factors. Through these aspects of socioscientific orientations, we intend to document changes in learners' perspectives on these dimensions resulting from their engagement in our place-based SSI course.

3.2 Implementation

Fifteen senior preservice science teachers (PSTs), consisting of 2 males and 13 females, participated in the place-based SSI instruction-related course being offered at a large state university in Ankara, Türkiye. The coursework was centered on the PST's concerns about local issues considering their SSI perspective in order to structure the dimensions of socioscientific orientations described in the previous section (see Herman et al., 2021) during the duration of the course. For this purpose, the instructors integrated Ankara's air pollution problem into the instructional context. As referenced in Table 3.1, the course was structured into seven sessions, including course tasks, field experience, and post-field experience components (see Table 3.1). Due to the Gaziantep-Kahramanmaraş earthquakes that occurred before the beginning of the Spring 2023 semester, the Turkish government decided to switch to hybrid education in universities, including distance education and face-to-face education. In the lessons during the hybrid teaching process, the interview forms were applied online via Google Forms as part of distance education. In addition, activities based on field experience were carried out face-to-face.

Table 3.1 Ankara's air pollution-focused place-based SSI course structure

Section of course	Course activity
Course tasks	*Session 1:* Air Pollution Socioscientific Issue View Form (Pre-test)
	Session 2: The Air Pollution Problem in Ankara Questionnaire
	Session 3: Preventing Air Pollution in Ankara View Form
	Session 7: Air Pollution Socioscientific Issue View Form (Post-test)
Field experience	*Session 4:* The Republic of Türkiye Ministry of Environment, Urbanization and Climate Change Gölbaşı Campus Field Trip
	Session 6: Self-reports for Interviews with Stakeholders Regarding Air Pollution in Ankara
Postfield work	*Session 5:* The Republic of Türkiye Ministry of Environment, Urbanization and Climate Change Gölbaşı Campus Field Trip View Form

On the first day of the course, the PSTs filled out a pre-test entitled Air Pollution Socioscientific Issue View Form. It aimed to examine their prior knowledge, thoughts, and perspectives relevant to the air pollution problem, addressing each dimension of socioscientific orientations.

In the initial part of the course, the instructors provided a basic knowledge of air pollution and mentioned air pollution in Ankara as place-based SSI instruction. Air pollution is considered one of the most prominent environmental issues in Ankara, and it reaches levels that can threaten public health concerns from time to time (Yavuz, 2020). During this learning process, the preservice teachers were informed regarding common air pollutants that were called particulate matter (PM), ozone (O_3), nitrogen dioxide (NO_2), carbon monoxide (CO), sulfur dioxide (SO_2) and their health effects under the title of air quality. The instructors also provided handouts and presented PowerPoints on how these air pollutants occur and which parameters (traffic, heating, industry, etc.) play important roles.

Furthermore, the PSTs were provided with the Ankara City Clean Air Action Plan Manual, published by The Republic of Türkiye Ministry of Environment, Urbanization and Climate Change and Ankara Metropolitan Municipality. This manual included air pollutant measurement values, particularly $PM_{2.5}$ and PM_{10} index values, obtained from each air Emission Measurement Center (EMC) located in the province of Ankara since 2020. The manual emphasized precautionary recommendations and encouraged students to contemplate measures for preventing air pollution in Ankara. This course component aimed to facilitate the development of PSTs' views on each dimension of socioscientific orientations, particularly regarding Scientific Evidence Views.

In the following course sessions, the instructors administered the Air Pollution Problem in Ankara Questionnaire and the Preventing Air Pollution in Ankara View Form. Finally, before the semester ended, the preservice teachers completed the Air Pollution Socioscientific Issue View Form once again as a post-test. These assessments allowed the PSTs to evaluate their socioscientific orientations while actively engaging with a local problem.

As part of the field experience component, the instructors organized a field trip for the PSTs to visit the Republic of Türkiye Ministry of Environment, Urbanization and Climate Change Gölbaşı Campus in Ankara. During this trip, the PSTs had the opportunity to enhance their place-based SSI perspective by observing the air pollution control and research laboratory with guidance from field experts. Additionally, a meeting was arranged with the Director of Clean Air for Northern Central Anatolia, who is responsible for overseeing air quality in Ankara and other provinces such as Bartın, Bolu, Çankırı, Düzce, Eskişehir, Karabük, Kastamonu, Kırıkkale, Kırşehir, Kütahya, Yozgat, and Zonguldak. The director delivered an interactive presentation and shared the latest air quality emission measurement reports with the PSTs. The field trip also included a visit to air control measurement analyzers (see Fig. 3.1).

In the subsequent stage of the field experience, the PSTs conducted interviews with various stakeholders who have been directly affected by air pollution in Ankara. These stakeholders included a coal dealer, an industrialist, a sugar factory worker, a mechanical engineer, an iron factory manager, a furniture maker, and local residents living in areas with poor air quality. The PSTs documented these interviews as self-reports, which provided valuable insights into the perspectives and experiences of individuals impacted by air pollution.

During the post-field work phase, the PSTs completed The Republic of Türkiye Ministry of Environment, Urbanization and Climate Change Gölbaşı Campus Field Trip View Form. They shared their field experiences in relation to socioscientific orientations. This form also prompted the PSTs to critically evaluate the objectivity and realism of the declaration made by the Director of Clean Air for Northern Central Anatolia regarding Ankara's air pollution problem. Furthermore, the form allowed the PSTs to describe the extent to which they evaluated the parameters and information concerning air pollution in Ankara from their socioscientific perspective.

These activities aimed to provide the PSTs with a comprehensive understanding of the socioscientific dimensions of the air pollution issue in Ankara and encourage critical analysis and reflection on various stakeholders' perspectives.

3.2.1 Data Sources

In order to assess the potential changes in PSTs' socioscientific orientations, we administered a questionnaire consisting of multiple-choice questions accompanied by open-ended explanations. The questionnaire was designed based on the framework of socioscientific orientations outlined in the study by Herman et al. (2021). It comprised eight multiple-choice questions, each followed by a section where PSTs were encouraged to provide a rationale for their chosen response. The multiple-choice options for each question presented two or three arguments representing different perspectives on the respective aspects of their socioscientific orientations. For instance, regarding the aspect of feeling responsibility, the PSTs were provided with the following options to choose from:

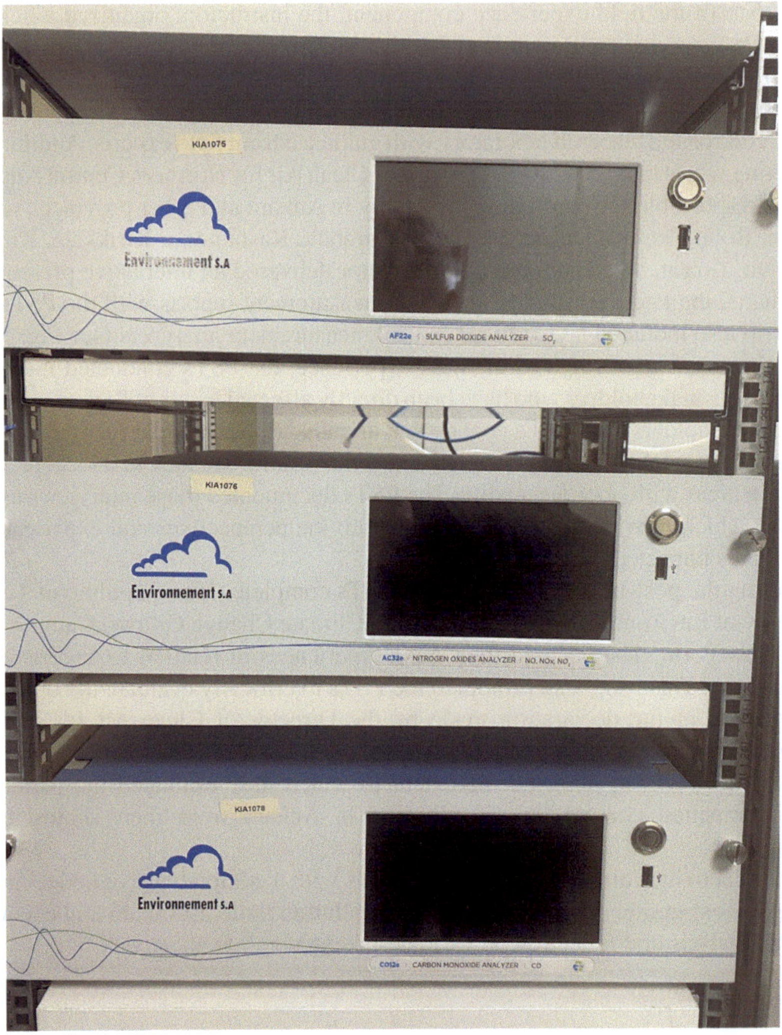

Fig. 3.1 The air pollution control and research laboratory

- I believe that the responsibility for the formation of the air pollution problem lies primarily with large companies and institutions that lack adequate prevention measures. (Naive view)
- Although I do not consider myself personally responsible for the air pollution problem, I believe that voluntary contributions from individuals like me are important in finding a solution. (Moderate view)
- I hold the belief that everyone in our society contributes, directly or indirectly, to the emergence of the air pollution problem. Therefore, I feel that everyone should shoulder responsibility and contribute, regardless of the challenges involved. (Informed view)

To ensure a comprehensive understanding of PSTs' socioscientific orientations, we utilized a questionnaire designed with specific options to cater to their relatively limited experience in engaging with socioscientific issues. This approach aimed to gather a more accurate depiction of their perspectives.

Additionally, we incorporated coursework assignments as another valuable data source. One notable assignment was the culminating activity administered at the end of the course. For this task, PSTs were tasked with evaluating a policy proposal to reduce air pollution in Ankara. The proposal outlined various measures, such as increased taxation on personal car usage, burning of coal and wood for heating, and additional charges imposed on industrialists based on their air pollution levels. PSTs were asked to express their support or opposition to the policy proposal and evaluate its potential for success. By including this assignment, we aimed to gain further insights into PSTs' socioscientific orientations and their ability to critically analyze real-world environmental policies.

3.2.2 Outcomes

To provide a comprehensive summary of the changes in socioscientific orientations among the 15 PSTs, we analyzed and compared their pre and post-responses to the questionnaire. By tabulating the data, we were able to present a general overview of the shifts and developments in their perspectives throughout the course. This approach allowed us to track the evolution of their socioscientific orientations and gain insights into the impact of the instructional interventions on their attitudes and beliefs regarding air pollution and related issues.

Upon initial examination of Table 3.2, it becomes evident that most of the PSTs had an informed view of various aspects of their socioscientific orientations at the beginning of the course. However, a closer analysis of their explanations reveals a different narrative. Despite selecting options aligned with informed views, the justifications provided by the PSTs did not align with their choices. This finding suggests a potential inconsistency or lack of coherence in their understanding and articulation of SSI. Further exploration of their explanations and a deeper investigation into their reasoning may provide valuable insights into this discrepancy and shed light on the underlying factors influencing their socioscientific orientations. For instance, for the perspective-taking aspect of the questionnaire, the majority of PSTs chose the following option

> I think that the solution to the air pollution problem can only be achieved with the joint contribution of all parties to the problem (state, municipality, industrialists, citizens, etc.). (Informed View for Perspective Taking Aspect)

A common explanation for this was that

> The problem does not only belong to a certain group, and only a certain group cannot solve it. (Meryem's Pre-test Perspective Taking Explanation)

Table 3.2 Changes in PSTs' socioscientific orientations from pre to post-participation in place-based SSI course

Aspects	Levels	Pre	Post	Change
Interconnectedness	Naive	4	5	1*
	Informed	8	10	2**
	Not represent	3	0	−3**
Sustainability	Naive	3	2	−1**
	Informed	12	13	1**
	Not represent	0	0	0
Perspective taking	Naive	0	0	0
	Informed	15	15	0
	Not represent	0	0	0
Moral and ethical sensitivity	Naive	3	1	−2**
	Informed	11	11	0
	Not represent	1	3	2*
Empathic concerns	Naive	7	8	1*
	Informed	5	7	2**
	Not represent	3	0	−3**
Feeling of responsibility	Naive	0	1	1*
	Informed	13	14	1**
	Not represent	2	0	−2**
Willingness to act	Naive	0	0	0
	Moderate	8	3	−5**
	Informed	7	12	5**
	Not represent	0	0	0
Affordances and constraints of scientific evidence	Naive	2	1	−1**
	Informed	12	14	2**
	Not represent	1	0	−1**

Note. "Not Represent" includes teachers who say that the given choices do not reflect their views and who write their own opinions

0 represent no changes, *represent a negative change, **represents a positive change towards informed socioscientific orientation

Notably, this PST believed involving more parties in the solution would lead to better outcomes. However, she seemed unaware of the potential conflicts that could arise from involving multiple parties and the difficulty of managing such conflicts to reach a consensus. Importantly, her reasoning did not undergo any significant changes throughout the course.

This pattern was consistently observed among most participants regarding various aspects, including sustainability, moral and ethical sensitivity, feeling of responsibility, and the affordances and constraints of scientific evidence. It suggests a common trend where PSTs may hold certain beliefs or preferences without fully considering the complexities and challenges associated with those perspectives. This indicates the need to further explore and develop their understanding in these areas. For instance, regarding affordance and constraints of scientific evidence, most of the PSTs chose the option

> I do not believe that the information obtained from scientific research studies alone will solve the air pollution problem. Air pollution is a complex problem with many variables (such as social, ethical, and economic). The solution of such problems is not only possible with scientific evidence. (Informed View for Affordances and Constraints of Science Evidence Aspect)

A common explanation for this informed view was that

> Scientific evidence strongly supports these issues, but alone is not sufficient. Since this issue is social, it should be evaluated together with many effects. (Hafize's Pre-test Affordances and Constraints of Scientific Evidence)

This PST recognizes the importance of considering social and moral aspects alongside scientific evidence in addressing the issue of air pollution in Ankara. However, she struggles to articulate how this integration can be achieved in practice clearly.

In general, most of the PSTs were initially inclined towards informed views of various aspects of socioscientific orientation. This inclination was further strengthened through their engagement in place-based course activities. However, it is worth noting that many of the PSTs' explanations did not delve deeply enough into the core aspects of socioscientific orientations.

Another notable trend observed in the data is the increase in informed views regarding the aspect of willingness to act. Initially, more than half of the PSTs expressed a moderate response to this aspect. The moderate response stated:

> I also think that those who have a share in the emergence of the air pollution problem are the ones who should take the most role and responsibility in solving the problem. Therefore, the causes of air pollution should be identified, and the responsibility for the solution should be assigned to them. However, I also think that I can personally contribute to my volunteering. (Moderate View of Willingness to Act)

For instance, one of the PSTs explained her choice of moderate view by stating *"As individuals, of course, we should take responsibility, but larger masses have to take responsibility and raise awareness on this issue." (Hilal Pre-test Willingness to Act).*

Indeed, the PST's explanation indicates that she did not assign personal responsibility to herself regarding the issue of air pollution in Ankara. Her perspective was that individuals' responsibility in addressing the problem could be voluntary. This type of explanation was commonly observed among those who chose the moderate view. Her response after the course changed to an informed view that states:

> I am aware that I am as responsible as everyone else in the formation and solution of air pollution. For this reason, I know that you should change my behavior, as well as everyone else, in solving the problem, although it will be difficult for me. (Informed View of Willingness to Act)

Her explanation was that

> More or less, every individual and institution in the society causes air pollution. Therefore, each stakeholder should contribute to the solution of the air pollution problem. (Hilal Post-test Willingness to Act)

It is encouraging to see that the PST's post-course response demonstrates a deeper understanding of the causes of air pollution and a recognition of the need for individuals to take voluntary responsibility and action to address the issue.

This indicates that the course had an impact on their awareness and willingness to make positive changes.

However, it is worth noting that despite the increase in informed views, the PSTs' explanations still lacked contextual links to the specific issue of air pollution in Ankara. Their rationales remained naive, context-free, and general in nature. This suggests that while their overall understanding improved, there is still room for further development in connecting their socioscientific orientations to the specific context of the problem they were studying.

3.3 Transferring Air Pollution SSI to the Global Concept in Science Education

Based on the latest IQAir (2022) world air quality reports, annual average $PM_{2.5}$ concentrations (μg/m^3) of many metropolitan cities around the world have air quality problems similar to Ankara. For instance, the annual average of $PM_{2.5}$ concentrations is 29.8 μg/m^3 for Beijing-China, 36.2 μg/m^3 for Jakarta-Indonesia, 47.4 μg/m^3 for Cairo-Egypt, and 92.6 μg/m^3 for Delhi, India.

Placing SSIs, such as air pollution, within the context of everyday concerns of a community is indeed crucial for engaging students and fostering their understanding of global issues. By addressing local or place-based issues, students can develop a deeper connection and relevance to the topic. It's encouraging to see that many international science curricula incorporate air pollution under the unit of humans and the environment, recognizing the interplay between human activities and the environment. This approach allows students to explore the causes, effects, and potential solutions to environmental pollution, including air pollution. Similarly, aligning the lesson tasks and activities related to Ankara's air pollution problem with the Human and Environment Relationship unit in the Turkish Science Education Curriculum for primary education helps create a comprehensive understanding of the topic. By exploring the effects of human activities on the environment, students can develop a sense of responsibility and actively contribute to sustainable practices.

Overall, the integration of place-based issues and the emphasis on the human-environment relationship in science education curricula provide students with valuable knowledge and skills to address environmental challenges, including air pollution, both locally and globally.

The Science Education in Europe (SSE) (2012) benchmarks highlighting air pollution as a result of industry and pollution align with the concept of environmental sustainability. By emphasizing the impact of human activities on the environment, these benchmarks emphasize the need to address air pollution as part of sustainable practices. In the case of the American NGSS (2013), including air pollution from burning fossil fuels under the science units of Earth and Human Activity and Energy indicates a comprehensive approach to understanding the causes and consequences of air pollution.

According to the literature on undergraduate students' views on air pollution as an SSI in Eastern countries like Indonesia and Türkiye, the focus is often on observation, inquiry-based thinking, and experimentation. By incorporating laboratory instructions related to industrial air pollution and transportation issues, instructors aim to provide students with hands-on experiences that connect to real-world contexts, such as the Industrial Revolution and incomplete combustion (Karışan, 2014). Collaborative learning and argumentation, including role-playing scenarios and discussions, are favored to encourage students' engagement and development of socioscientific reasoning skills (Susilawati et al., 2021).

In Western countries such as England, Spain, Sweden, and the Netherlands, instructors emphasize evidence-based decision-making and informal reasoning skills in enhancing students' socioscientific reasoning about air pollution. PSTs are encouraged to design and implement SSI content-based lesson plans, allowing them to reflect on their learning process (Espeja & Couso, 2020). Case studies relevant to socioscientific inquiry-based learning activities are presented, providing PSTs with opportunities to discuss their experiences and deepen their understanding of air pollution as an SSI (Amos et al., 2020).

These varied instructional approaches reflect the diverse contexts and educational practices in different countries. By adopting strategies incorporating real-world connections, collaborative learning, argumentation, and evidence-based decision-making, instructors can empower students to develop a comprehensive understanding of air pollution and engage in meaningful discussions and actions to address this important environmental issue.

3.4 Conclusion and Implications

Our study provides valuable insights into the impact of engagement in contentious science-based societal issues-focused instruction (SSI) on PSTs' socioscientific orientations. The findings show that PSTs tended to choose informed views before their involvement in SSI-based instruction. While the instruction helped more PSTs choose informed views, their explanations remained naive, context-free, and general. This contrasts with previous studies that have reported positive impacts on various learning gains through place-based SSI instruction (Herman et al., 2021; Kim et al., 2020; Kinslow et al., 2019).

One possible explanation for this discrepancy is the degree of immersion in SSI instruction. In our study, the transition from face-to-face to online instruction at the beginning of the semester may have affected the sense of belonging and engagement in classroom activities, particularly for PSTs who were not able to participate in field trips and other activities. Additionally, the limitations of the online learning management system may have hindered effective interaction and guidance from the course instructor during activities.

Considering these factors, it becomes evident that the success of place-based SSI instruction relies heavily on the degree of learners' immersion in SSI and the level of guided interaction with the parties involved in the societal issue. Future studies could explore ways to enhance immersion and interaction in online learning environments, as well as identify effective strategies for guiding and supporting learners during SSI activities.

One significant finding from the SSI-based course was the increased willingness of PSTs to take action to mitigate air pollution in Ankara. Initially, they attributed the responsibility for preventing air pollution to companies and government organizations, with individuals having only voluntary responsibilities. However, as they developed a more informed understanding of the causes of air pollution in Ankara, they began to recognize the contributions of individuals to the issue. This shift in perception was evident in their responses to post-fieldwork forms.

When individuals feel a greater sense of responsibility for an issue's emergence, it often translates into a stronger willingness to take action toward its resolution. In the case of air pollution in Ankara, the increased understanding and awareness among the PSTs led to a more intense motivation to actively contribute to addressing the problem. This finding highlights the importance of providing learners with opportunities to develop a deeper understanding of societal issues, as it can foster a sense of personal responsibility and promote a greater willingness to take action.

The assessment of learners' socioscientific orientations is an area that warrants further discussion. In our study, we utilized a questionnaire adapted from Herman et al.' (2021) framework of socioscientific orientations to gauge the perspectives of our PSTs. To encourage more nuanced reflections, we provided choices that encompass both naive and informed views for each aspect. Prior to their engagement in SSI, the majority of PSTs exhibited a propensity towards choosing informed views, but their explanations remained predominantly naive. This suggests that context-free assessments with general statements may not sufficiently capture learners' socioscientific orientations. To achieve a more accurate documentation of these orientations, it is necessary to employ a range of assessment methods, such as interviews and contextually specific questions. However, implementing such comprehensive approaches may pose challenges for larger-scale studies. Therefore, further research is needed to develop refined assessment strategies that effectively capture learners' socioscientific orientations in diverse educational contexts.

References

Amos, R., Knippels, M., & Levinson, R. (2020). Socio-scientific inquiry-based learning: Possibilities and challenges for teacher education. In M. Evagorou, J. A. Nielsen, & J. Dillon (Eds.), *Science teacher education for responsible citizenship* (pp. 41–61). Springer.

Bencze, J. L., El Halwany, S., & Zouda, M. (2020). Critical and active public engagement in addressing socioscientific problems through science teacher education. In M. Evagorou, J. A. Nielsen, & J. Dillon (Eds.), *Science teacher education for responsible citizenship* (pp. 63–83). Springer.

Espeja, A. G., & Couso, D. (2020). Introducing model-based instruction for SSI teaching in primary pre-service teacher education. In M. Evagorou, J. A. Nielsen, & J. Dillon (Eds.), *Science teacher education for responsible citizenship* (pp. 153–171). Springer.

Fortus, D., Jing, L., Neumann, K., & Sadler, T. D. (2022). The role of affect in science literacy for all. *International Journal of Science Education, 44*(4), 535–555.

Herman, B. C., Newton, M. H., & Zeidler, D. L. (2021). Impact of place-based socioscientific issues instruction on students' contextualization of socioscientific orientations. *Science Education, 105*(4), 585–627.

Hodson, D. (2009). *Teaching and learning about science: Language, theories, methods, history, traditions and values*. Sense Publishers.

IQAir. (2022). *2022 world air quality report*. Goldach.

Karışan, D. (2014). *Exploration of preservice teachers' reflective judgment and argumentation skills revealed in a socioscientific issue-based inquiry laboratory course* (Thesis No. 377864) [Doctoral dissertation, The Graduate School of Social Sciences of Middle East Technical University].

Kim, G., Ko, Y., & Lee, H. (2020). The effects of community-based socio scientific issues program (SSI-COMM) on promoting students' sense of place and character as citizens. *International Journal of Science and Mathematics Education, 18*, 399–418.

Kinslow, A. T., Sadler, T. D., & Nguyen, H. T. (2019). Socio-scientific reasoning and environmental literacy in a field-based ecology class. *Environmental Education Research, 25*(3), 388–410.

Ministry of National Education (MEB). (2018). *Science curriculum (Primary and middle school 1, 2, 3, 4, 5, 6, 7 and 8th grades)*. Ankara.

NGSS Lead States. (2013). *Next generation science standards: For states, by states*. National Academies Press.

Powell, W. A., Newton, M. H., & Zeidler, D. L. (2021). Impact of socioscientific issues on middle school students' character and values for global citizenship. In W. Powell (Ed.), *Socioscientific issues-based instruction for scientific literacy development* (pp. 56–91). IGI Global.

Romine, W. L., Sadler, T. D., & Kinslow, A. T. (2017). Assessment of scientific literacy: Development and validation of the quantitative assessment of socio-scientific reasoning (QuASSR). *Journal of Research in Science Teaching, 54*, 274–295.

Sadler, T. D. (2004). Informal reasoning regarding socioscientific issues: A critical review of research. *Journal of Research in Science Teaching, 41*(5), 513–536.

Sadler, T. D., Barab, S., & Scott, B. (2007). What do students gain by engaging in socioscientific inquiry? *Research in Science Education, 37*(4), 371–391.

Science Education in Europe: National Policies, Practices and Research. (2012). *Science education in Europe: National policies, practices and research* (pp. 1–167). https://data.europa.eu/doi/10.2797/7170

Siarova, H., Sternadel, D., & Szőnyi, E. (2019). *Research for CULT committee—Science and scientific literacy as an educational challenge*. European Parliament, Policy Department for Structural and Cohesion Policies.

Susilawati, Paidi, A. N., & Irwanto, I. (2021). Socio-scientific issues as a vehicle to promote soft skills and environmental awareness. *European Journal of Educational Research, 10*(1), 161–174. https://doi.org/10.12973/eu-jer.10.1.161

The Organization for Economic Cooperation and Development (OECD). (2017). *PISA 2015 assessment and analytical framework: Science, reading, mathematics, financial literacy and collaborative problem solving*. OECD Publication Revised Edition.

The Organization for Economic Cooperation and Development (OECD). (2018). *The future of education and skills: Education 2030*. Position paper. http://www.oecd.org/education/2030/E2030%20Position%20Paper%20(05.04.2018).pdf

Yavuz, C. I. (2020). Kentsel alanda hava kirliliği: Hava kirliliği izleme ağı Ankara istasyonlarının beş yıllık PM_{10} ölçüm verilerinin incelenmesi [Urban air quality: Evaluation of five years PM_{10} measurement data in capital city of Turkey]. *Sağlık ve Toplum, 20*(3), 96–102.

Zangori, L., Ke, L., Sadler, T., & Peel, A. (2020). Exploring primary students' causal reasoning about ecosystems. *International Journal of Science Education, 42*(11), 1799–1817.

Zeidler, D. L., Herman, B. C., & Sadler, T. D. (2019). New directions in socioscientific issues research. *Disciplinary and Interdisciplinary Science Education Research, 1*, 1–9.

Chapter 4
Developing Argumentation Skills in Place-Based SSIs: Construction of a Nuclear Power Plant in Sinop

Nurhan Öztürk and Ayşe Yenilmez Türkoğlu

4.1 Introduction

Scientific literacy is often held up as the basic goal of science education all over the world recently. Scientifically literate individuals are expected to make scientific inquiries, claims, and warrants or justifications, interpret and evaluate evidence, and construct deeper understandings of science (Lin & Mintzes, 2010). Scientific knowledge is constructed through social interaction, negotiation, and communication in the context of specific discourse practices (Driver et al., 1994). Meaning is constructed when students interpret and reinterpret events by using their knowledge (Barnes, 1992). In achieving these goals, the aim of science education is defined as not only accomplishing scientific concepts but also learning and engaging in scientific discourse (Bricker & Bell, 2009). At this point, engaging in argumentation and developing argumentation skills are central to science education (Kuhn, 2010). Argumentation is identified as the epistemological basis of science (Driver et al., 1994), and it is defined as the language of science (Duschl et al., 1999) since it covers the process that scientists engage in as they attempt to reach consensus (Tippett, 2009). Argumentation includes the commitments that are used to support one's own argument (Walton, 1989). Advanced argumentation skills, however,

N. Öztürk (✉)
Department of Mathematics and Science Education, Faculty of Education,
Sinop University, Sinop, Türkiye
e-mail: nurhanozturk@sinop.edu.tr

A. Yenilmez Türkoğlu
Department of Mathematics and Science Education, Faculty of Education,
Alanya Alaaddin Keykubat University, Alanya, Antalya, Türkiye
e-mail: ayse.yenilmez@alanya.edu.tr

B. Namdar, E. Karahan (eds.), *Socioscientific Issues Focused Teacher Education*, Contemporary Trends and Issues in Science Education 60,
https://doi.org/10.1007/978-3-031-55233-5_4

cover the identification of the weaknesses in the opponent's argument and refute his position (Walton, 1989). The use of evidence is also defined as a key component of skilled argumentation (Kuhn et al., 2013). Students will justify their own positions and be better at engaging in discussions on controversial issues if they learn to construct valid scientific arguments and integrate their thinking skills with socioscientific knowledge (Lin & Mintzes, 2010). Studies, however, report that students' argumentation skills are underdeveloped (Sadler, 2004), and the understanding of how these skills can be developed is limited (Zohar, 2008). An urgent need to examine ways to support argumentation skills, therefore, is still a pressing issue (Iordanou & Constantinou, 2014). At this point, educational interventions can serve as alternatives to promote the development of argumentation skills. Research on argumentation literature points to the importance of authentic learning, the idea of a community of learners, the need for explicit instruction, and the existence of multiple opportunities and discourses for practicing argumentation skills (Tippett, 2009).

In science education, argumentation processes can occur in the discussions between the students and the teacher as students try to reason and find plausible explanations for phenomena observed during laboratory work or as they are discussing SSIs with their ethical and societal implications (Christenson, 2015). SSI-based learning experiences help students become better at argumentation since dealing with these issues requires understanding the nature of the issue and getting prepared to engage in decision-making and position-taking in the context of the issue. Typically, SSI-based learning experiences occur in formal learning environments like classrooms through engaging in hypothetical scenarios. Students make *hypothetical decisions* in these situations and make so-called opinions about issues they do not actually adopt or take part in. These shortages inhibit efforts to become knowledgeable and feel responsible towards SSIs. Students may perceive hypothetical SSI scenarios as detached from their daily lives, and learning through these experiences will likely be poor since they will not interact with people and environments affected by SSIs (Zeidler et al., 2019). Demonstrating sociocultural awareness and moral sensitivity in developing perspectives and making decisions about SSIs in these cases can also be problematic (Zeidler et al., 2019). Decision-making in particular real-world contexts, at this point, emerges as a necessity.

For students to conceptualize and internalize SSIs, the contextual features of the learning environment, like the cultural milieu and the stakeholders impacted by SSI, are crucial (Zeidler et al., 2019). At this point, place-based SSI instruction provides first-hand experiences with diverse stakeholders like local communities, workers, authorities, and suppliers, who have knowledge and deep perspectives about local SSIs. Research shows that place-based SSI instruction develops students' knowledge, contextualized NOS understanding, compassion for others impacted by SSI, and willingness to resolve SSIs (Herman, 2018). Students realize and draw attention to the *real* problems of their region (Çapkınoğlu & Yılmaz, 2018), and they become more knowledgeable about the local SSIs (Demircioğlu & Uçar, 2014). Their multidimensional perspectives and argumentation skills are also improved by empirically addressing a local issue (Demircioğlu & Uçar, 2014; Lee & Grace, 2010). Therefore, sociocultural awareness and moral sensitivity are developed through place-based SSI interactions (Zeidler et al., 2019). As a matter of fact, understanding the beliefs, attitudes, and behaviors of local communities affected by local issues and acting accordingly are among the fundamentals of SSI instruction and science education, as well.

Students' experience with local SSIs through making direct contact with the society they live in is essential. However, how can we expect teachers to support the development of students' awareness, sensitivity, or argumentation skills if they haven't developed these skills adequately themselves? Local SSIs, actually, concern, either directly or indirectly, not only a certain segment of the region but also all occupational groups, all residents from seven to seventy, and are usually on everyone's agenda. In this respect, the existence of individuals who are aware of the issue, who make correct statements about the issue in the context of science literacy, who produce arguments supported by scientific data, and who share this knowledge and experience with their environment provides a tendency to the local issue, invite them to be sensitive, and are aware of their role as global citizens is important. At this point, discussing local SSIs with PSTs (and teachers, as well) currently living in the region will certainly be effective in raising their awareness of the issue, improving the quality of their arguments, and making conscious decisions by being aware of the issue. PSTs, who will be prepared for the profession and will transfer all their professional knowledge to large groups, will contribute to their ability to look at social problems in a multi-faceted way if they have the knowledge about local issues, develop arguments, live an effective decision-making process, and experience an expert-supported teaching program in teaching local issues. In this direction, it is thought that an expert-supported teaching program will be an application that will support the development of preservice teachers' awareness of the local issue and the development of their argumentation skills. The implementation outputs are thought to be important in giving ideas to preservice teachers, teachers, and researchers on how local issues can be used in classroom practices to develop argumentation skills.

4.2 Implementation

4.2.1 SSI Context: Construction of a Nuclear Power Plant in Sinop

Industrial and economic growth, digitalization, and the growing population need in both developing and developed countries all over the world create an increasing need for safe, continuous, and sustainable energy production. Most countries depend on fossil fuel energy sources, such as natural gas, oil, and coal, to fulfill their energy needs, but the emissions of greenhouse gasses and pollutants from the use of fossil fuels lead to significant environmental problems and the deterioration of human health. Such negative impacts and the ongoing energy dependency drive the need for nuclear power as being the only large-scale source of greenhouse gas-free energy. Established in 1954 as the first power reactor for energy production in Moscow, nuclear energy became a commercially viable energy source in the 1960s, and electricity production from nuclear energy increased rapidly after the 1970s due to the oil crisis. Although a slowdown in production occurred thereafter as a result of the entry of natural gas into the energy market and the accidents in Three Mile Island in 1979 and in Chernobyl in 1986, a continuous increase in the production of total nuclear power is seen over the years (Eral, 2015). Nuclear energy today generates around 11% of total global electricity, and this amounts to one-third of the low-carbon electricity produced in the world (Mathew, 2022).

Like many other developing countries, Türkiye's energy demand is also on the rise, ranking first in Europe and second in the world after China, constantly increasing at an annual average rate of 7–8% (Eral, 2015). In Türkiye, 34.6% of electricity production comes from coal, 22.2% from natural gas, 20.6% from hydraulic energy, 10.8% from wind, 4.7% from the sun, 3.3% from geothermal energy, and 3.7% from other sources (Ministry of Energy and Natural Resources [MENR], 2023). By 2023, the number of electrical power generation plants in the country is 11,845; 751 of which are hydroelectric, 67 are coal, 361 are wind, 63 are geothermal, 346 are natural gas, 9767 are solar, and 490 are other sourced power plants (MENR, 2023). Gathering energy from several sources mentioned here, the country does not have an NPP yet, but it seems that the NPP construction attempts have been going on for a while.

In Türkiye, the Atomic Energy Commission (Atom Enerjisi Komisyonu [AEK]) was established in 1956 as the first significant step in nuclear energy. The first studies on the construction of nuclear power plants were initiated by the Commission in 1965. In the 1970s, the construction of a reactor to generate electricity was planned but could not be realized. In 1982, AEK was reorganized as Turkish Atomic Energy Agency (Türkiye Atom Enerjisi Kurumu [TAEK]), and bids were taken from Atomic Energy Canada Ltd (AECL), Siemens & Kraft Werk Union (KWU) and General Electric (GE) for the construction of four nuclear power plants, two units in Akkuyu and two units in Sinop (Eral, 2015). The agreement was reached with AECL and Siemens-KWU in 1984; however, the companies withdrew from the negotiations due to the loan guarantee problems and the risk of the Build-Operate-Transfer model (Eral, 2015). Various other negotiations took place in the following years, but no agreement was reached. In 2004, the Ministry of Energy and Natural Resources and TAEK announced the construction of power plants, which will begin in 2007, and the commission of the first unit in 2012. TAEK announced the studies for the construction of Sinop Nuclear Technology Center (Sinop Nükleer Teknoloji Merkezi [SNTM]) in the same year where research reactors, fuel cycle facilities, support facilities, nuclear technology research and development facilities, and training facilities would be carried out. Following the signing of the Cooperation Agreement on the Establishment and Operation of Nuclear Power Plants in the Akkuyu Field between the governments of Türkiye and Russia in 2010, the foundation of the Akkuyu NPP offshore structures was laid in 2015. While the construction process of Akkuyu NPP continues, the Sinop Nuclear Power Plant (Sinop NPP) installation process also started in 2013. The construction of Sinop NPP was planned to start in 2017, and the commissioning years of the units were foreseen as 2023 for Unit I, 2024 for Unit II, 2027 for Unit III, and 2028 for Unit IV (Eral, 2015).

Nuclear energy for the country is of great importance in terms of ensuring energy supply security, reducing the dependence on foreign energy, increasing resource diversity, providing continuous energy availability and competitive electrical energy prices, and reducing the country's current account deficit. It also meets environmental sensitivities by preventing greenhouse gas emissions and climate change. The problem of nuclear waste, the health effects of nuclear radiation, and especially the possibility of a nuclear accident, however, raise questions about the future of nuclear energy in the country, as well as in the world. Factors like energy demand, carbon emission targets, economic competition with fossil fuels, environmental sensitivities, and, more significantly, the *public attitude* will determine the future of nuclear energy in Türkiye.

Sinop is a settlement that stands at the far north of Türkiye with its historical and natural beauties. The city is a natural harbor surrounded by the Black Sea on three sides. Along with its various endemic plants, it has a rich forest and vegetation. With its rich biological diversity, the Sarıkum Nature Reserve area in Sinop offers breeding and accommodation opportunities for different species of birds at all times of the year and attracts the attention of researchers and visitors. Fishing activities are also quite common in the city. Sinop is on the migration route of species with high economic value, such as bonito, bluefish, and anchovy, and deep fish, such as turbot, haddock, and red mullet, are hunted abundantly. Fishing activities in the city play an important role in meeting the country's seafood needs, as well.

However, constructing the nuclear power plant in the city is a controversial issue in Sinop. Various supporting or opposing statements from scientific communities, scholars, decision-makers, non-governmental organizations (NGOs), and media are on the society's agenda. One of the most common criticisms of the construction of the NPP is the possibility of accidents and the devastating consequences of these accidents (Ateş & Saraçoğlu, 2013; Çelik et al., 2015; Eş et al., 2016; Kılınç et al., 2013). Citizens also refer to the health problems caused by nuclear waste and radioactivity (Akyüz, 2017; Ateş & Saraçoğlu, 2013; Es & Yenilmez Türkoglu, 2021; Yenilmez Türkoğlu & Öztürk, 2019; Yıldırım & Örnek, 2007), and to the rarity of the raw material for nuclear fuel (Çelik et al., 2015). They also think that the construction of an NPP will affect tourism activities negatively since hesitation towards visiting the region will arise among individuals (Ateş & Saraçoğlu, 2013; Yıldırım & Örnek, 2007; Yılmaz, 2015). Supporters, on the other hand, claim that nuclear energy will decrease Türkiye's energy dependency on other countries (Akyüz, 2017; Ateş & Saraçoğlu, 2013; Eş et al., 2016; Es & Yenilmez Türkoglu, 2021; Yenilmez Türkoğlu & Öztürk, 2019), provide economic power (Yenilmez Türkoğlu & Öztürk, 2019) and it will reduce CO_2 gas emissions and therefore global warming and climate change (Yıldırım & Örnek, 2007). They also share the idea that having insufficient renewable energy sources in Türkiye, nuclear energy is a good alternative with high energy efficiency, significant return on cost, and a low amount of waste (Yılmaz, 2015). Either supporting or opposing, citizens' approaches to the NPP base on *economic* (Eş et al., 2016; Es & Yenilmez Türkoglu, 2021; Eş & Varol, 2019;

Ozturk & Bozkurt Altan, 2019; Türköz & Öztürk, 2020; Evren Yapıcıoğlu & Kaptan, 2018), *political* (Eş et al., 2016; Es & Yenilmez Türkoglu, 2021; Eş & Varol, 2019; Ozturk & Bozkurt Altan, 2019; Türköz & Öztürk, 2020), and *environmental* (Eş et al., 2016; Es & Yenilmez Türkoglu, 2021; Ozturk & Bozkurt Altan, 2019; Türköz & Öztürk, 2020) grounds.

With its impact on the economy, politics, and environment, the issue is still up-to-date and is constantly brought to the agenda with different perspectives. The issue is local, but due to its possible effects, it is also in the field of interest of all countries on a regional or even global scale. Residents' knowledge and arguments form the multi-faceted structuring of their perspectives on the issue. Along with the residents, employees, stakeholders, farmers, fishermen, and all other students of Sinop, preservice teachers studying at Sinop University are also interested in the NPP being constructed in this region. The issue is frequently addressed within the scope of various courses at the university, and it is important to determine and develop university students' understanding of the issue. In developing these thoughts, McClennen's (2016) five dimensions, communication and relationship building for place-based education, a partnership between educational institutions and society, students' inquiry and design, student-centered learning process, and interdisciplinary learning experiences, are considered as important. Thus, establishing communication between the authorities/experts living in the region (or having a relationship with the region) and the university students and enabling them to comprehend the local issue by participating in this process seems valuable. In this respect, it is thought that by discussing such an issue with the preservice teachers living there, the awareness of the issue and the development of the quality of their arguments can be ensured so that a more qualified and meaningful decision-making process on such a local SSI will be utilized.

4.2.2 Implementation Procedure

The study was conducted with the participation of ten preservice science teachers (PSTs) (nine females, one male) who were attending the third year of the Science Education program at Sinop University. An expert-supported online teaching process was carried out with them, and their awareness and quality of arguments were examined after the implementation. The implementation included theoretical information and discussions on SSIs and nuclear energy in general and the construction of a nuclear power plant in Sinop. Meetings were moderated by the researchers of this study, and presentations and discussions were held with the participation of four experts/researchers in these fields. The implementation procedure is presented in Fig. 4.1.

The implementation procedure took a total of 6 weeks, with 2 hours a week. A theoretical presentation on SSIs was done in the first week. PSTs already had experience with argumentation from their prior courses, but a short activity about writing

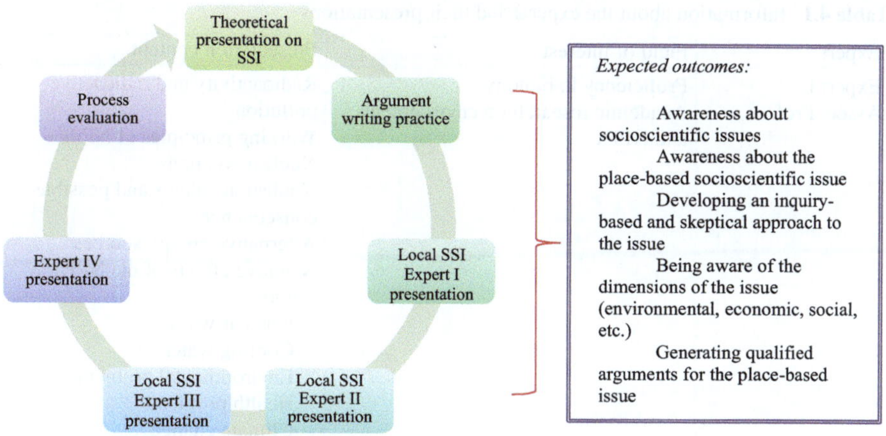

Fig. 4.1 The implementation procedure

arguments about SSIs was done after the presentation. PSTs' understandings and profiles about NPPs were also identified this week. During the following 4 weeks of the implementation, experts with differing backgrounds made presentations about nuclear energy and reflected the two opposing views about the NPP construction in Sinop. Information about the experts and their presentations is given in Table 4.1.

As Table 4.1 shows, all experts hold at least a Ph.D. degree and are related in some way to the issue of NPPs and/or the NPP planned to be constructed in Sinop. Expert I, who lives in Sinop and works at Sinop University, does not support the construction of an NPP in Sinop. Having a profession in biology, she focused mostly on the effects of a possible explosion on the environment and living things, sea life, and climate. She pointed to alternative energy sources to meet the energy demand. Expert II, who again lives in Sinop and works at Sinop University, supports the construction of an NPP in Sinop. He informed PSTs about the working principles of NPPs and discussed the misconceptions about energy production and nuclear waste storage from the perspective of a physics professional. He also emphasized the number of NPPs worldwide and discussed the countries that own and actively use NPPs. Expert III, who had a position in Türkiye's nuclear program, supports the construction of NPPs, as well. As an expert who had a senior position in Akkuyu NPP, he emphasized the positive aspects of the NPP for Sinop province by providing supporting data. In particular, he drew attention to the main topics related to the use of nuclear for security and military purposes, its contribution to the economy, new job opportunities and opportunities for suppliers, ensuring energy security, formation of new industries, and lower carbon dioxide emission. He also pointed to the possible risks of NPPs and their solutions, as well. The last expert, Expert IV, whose hometown is Sinop and who has an active position in Sinop Anti-Nuclear Platform, does not support the construction of an NPP in Sinop, as he takes an active role in Sinop's anti-nuclear platform. In his presentation, he mainly focused on Sinop NPP as a project and pointed to the problems with the features and evaluation of this

Table 4.1 Information about the experts and their presentations

Expert	Field of interest	Presentation content
Expert I: Assoc. Prof.	Proficiency in biology Academic research on environment and NPPs	Radioactivity and radioactive pollution Working principles of NPPs Nuclear weapons Nuclear accidents and possible consequences Alternative energy sources Negative effects of the NPP on Sinop Nuclear waste Cooling water Environmental problems Health problems Climate change Objections to the NPP in Sinop
Expert II: Prof.	Proficiency in physics Academic research on nuclear energy systems	Electricity Faraday's Law The structure of the atom and Uranium Fission, fusion reactions Nuclear fission Landscape properties for the construction of NPPs Working principles of NPPs Fuel efficiency of NPPs Energy production cost and efficiency Waste storage and solutions NNPs and nuclear energy in the world
Expert III: Research associate with a Ph.D.	Proficiency in environmental engineering and international relations Work experience in Akkuyu NPP	Project area in Akkuyu NPP Characteristics of reactors Nuclear agreements & responsibilities between countries Countries having NPPs Advantages of NPPs Use for security and military purposes Contribution to the economy Job opportunities Opportunities for suppliers Carbon dioxide emission Ensuring energy security Formation of new industries Training experts Possible risks and risk management Türkiye's nuclear program

(continued)

Table 4.1 (continued)

Expert	Field of interest	Presentation content
Expert IV: Prof.	Proficiency in economics Active position in "Anti-nuclear platform"	Projects and general project evaluation criteria Sinop NPP as a project Cost-benefit analysis Public interest Environmental impact assessment (EIA) report Expert opinions Cooling water problem Worker safety Public health Alternative energy sources and energy production rates North Anatolian Development Agency Plan

specific project. He presented data from the Environmental Impact Assessment Report written for Sinop NPP and indicated the problematic or missing issues in the report.

PSTs expressed their individual views and directed their questions to the experts during and after the presentations. Experts replied to the questions and clarified PSTs' questions. After the meetings with the experts ended, PSTs' understandings and arguments on the issue were taken each week via reflection papers—the implementation process ended by taking PSTs' general evaluation of the expert-supported teaching program.

4.2.3 *Outcomes*

An expert-supported teaching program was implemented for the purpose of developing PSTs' understanding and argumentation skills about NPPs through online meetings that provided opportunities to see professionals' explanations, interpretations, and experiences about the issue. It was seen before the implementation that PSTs had a limited understanding of NPPs, although NPP construction is a local SSI for them. While PSTs held supporting or opposing views, and most were undecided about the construction of an NPP in Sinop, it was seen that there were no undecided PSTs at the end of the implementation, and they were able to consider the local issue from wider perspectives. One of the PSTs, for example, pointed out the country's economy, the energy needs, and the job opportunities in Sinop and said:

> It [NPP] does not harm the environment as much as it is thought because greenhouse gas emission rates are lower when compared to the use of fossil fuels. It does not cause SO_2 emissions and, thus, acid rain. It will benefit the country economically. Not having high operating costs, nuclear power plants will meet a part of our country's energy needs, thus

> reducing our energy usage cost and foreign dependency. It also provides employment. It will take Türkiye one step further scientifically and technologically since nuclear technology provides scientific and technological progress.

In his opinion, he evaluates the issue from different dimensions, including environment, economy, science, politics, and technology. PSTs made evaluations with more than one dimension and used some data to strengthen their justifications. Another PST evaluated the issue from both local and global perspectives and said:

> While local issues sometimes seem to concern only that region, these issues have global or regional implications. This is especially true for nuclear energy. For instance, when we are talking about the risk of an explosion, in such a case, not only Sinop, Türkiye, and other countries will be adversely affected, just like in the Chernobyl accident. Another example is that, with this power plant, the energy needs of the whole of Türkiye will be met, so it ceases to be an issue specific to Sinop alone. So people from every region are participating in this discussion.

Traces from experts' opinions, explanations, or data from their presentations were also seen in PSTs' statements. It was determined that almost all PSTs stated that expert presentations and explanations were effective in their decisions on the issue. Therefore, PSTs' views and justifications on the issue were diversified and integrated with supporting and rebuttal argument elements. In the following statement of one of the PSTs, it is understood that he approached the issue from the parties that had an impact on him and that the expert opinions were effective in his decision:

> I listened to expert opinions. After all, they are the people who have worked in this field. I tried to think more realistically in my decision on the issue. It is obvious that the energy needs will increase in the future. I think nuclear power plants are more feasible for several reasons, such as the cost of maintaining renewable energy sources, considering alternatives, creating some obstacles, and having a shorter lifespan. With that in mind, I was struck by the country's current and probable future situation.

The most frequently cited topics from the presentations of the experts that are seen in the statements of the PSTs included the working principle of NPPs, reasons for constructing an NPP in Sinop, partner countries and agreements, storage of nuclear waste, energy efficiency in comparison to other energy sources, job opportunities, contribution to economy, harm to the ecosystem and health, cause and possible consequences of an accident, ways to reduce accidental risks, effects on sea-life, countries that use and stop to use NPPs, operation of the plant, positive effects on environment and air pollution, EIA report, public benefit, and amount of cooling water used.

PSTs also reflected the knowledge they obtained from the experts in their arguments. It was determined that PSTs, who presented only a claim and a brief justification before they were involved in the current implementation, used claims, data, justifications, backings, and rebuttals in their arguments after the implementation process. Items from the presentations of the experts were also observed in the arguments. They evaluated the issue from the viewpoint of those with the opposite view, and they especially benefited from scientific data and expert references in persuading the opposing view. They also cited experts' opinions that influenced them as they made decisions. For example, in his statement, the PST drew attention to the situation that affected his decision:

> Expert III's work experience at the Akkuyu nuclear power plant has greatly influenced my view. Because he has experience and is actively involved, his data about energy needs, nuclear power plants, and energy production influenced my mind.

While PSTs were developing arguments, they also approached the issue from the views of those who did not agree with them and emphasized the reason for which the opposing view defended itself and the ways of persuading the opposing view. For example, one of the PSTs does not want an NPP to be built in Sinop. However, it was understood that he was aware of opposing views from the statement: "*Of course, he may be right in terms of economy and foreign dependency…*". Another PST who does not support the construction of an NPP in Sinop tries to persuade others to express his argument as follows. It is seen from the statement that while convincing a person who has a different opinion from him, the PST looks at the issue in a multi-dimensional way and makes a statement by supporting it with some data.

> Alternative sources can, of course, be used instead of nuclear energy. However, considering each dimension, we must also consider the country's needs. Considering that our energy needs are increasing with the population… When we look at it from different angles, like cost, policy, and technology, we should choose the most reasonable energy source according to the current and probable situation. The most suitable choice for this situation would be nuclear energy. In this way, more energy is obtained in proportion, causing less greenhouse gas emissions compared to non-renewable energy sources that are currently concentrated.

PSTs reported the working principle of the power plant, the way the nuclear accidents happen, the criteria for constructing an NPP, international partners, the importance of the security container in an NPP, the effects of cooling water, and the EIA report as being the most interesting topics in the presentations of experts. Almost all PSTs said that they found answers to their questions during the meetings with experts. Only one PST stated that he could not find a clear answer to the question of whether cooling water has an effect on living things or not. PSTs stated that the situation that makes NPPs controversial is mostly due to the environmental and economic dimensions. They emphasized that discussions took place in the media with topics such as explosions, accidents and possible consequences, and energy deficit. Regardless of their approach to constructing an NPP, PSTs shared the common idea that it is important to take the opinions of the people living in that region while planning the construction of an NPP in Sinop.

PSTs stated that connecting with experts in SSI teaching experiences was effective in their decision-making process since they are specifically working and doing research in that field and they have one-to-one experience with the issue, making the information they provide more reliable. PSTs reported conducting a decision-making process by looking at the issue holistically with the information they received from each expert. For example, with the statement: "*I think Expert IV is very knowledgeable about the issue because he explained it with supporting data. He explained it in more detail, it sounded more reliable, and that was instrumental in shaping my ideas,*" it can be seen that the PST draws attention to the fact that information supported with data and detailed explanations are more effective and reliable. In another example, a similar thought is seen:

> Expert III gave information from right inside the field. Expert IV also gave information about the reports. I was also impressed by Expert II in terms of his characteristics. He did his presentation so enthusiastically, and his energy, style of expression, and willingness impressed me. I was a supporter of the construction of the NPP. However, I was also impressed by Expert I's presentation since she considered it from an environmental point of view. Expert III also impressed me because he was from the field and he was in the business.

PSTs seemed to be impressed by some characteristics of the experts. For example, they declared that Expert I is especially well-equipped in terms of environment. They found Expert II enthusiastic and excited about the issue and liked his data-supported approach to the issue. Expert III impressed them in terms of his working experience at Akkuyu NPP as a professional and, again, his data-supported presentation. Finally, Expert IV's explanations about the EIA report, his role in the Sinop anti-nuclear platform, and especially Sinop, being his hometown, were impressive on PSTs' decisions. As seen, being knowledgeable and presenting data arose as important factors for PSTs. Interestingly, however, being a resident of the city, or more specifically, the city being the hometown of the expert, was also considered a factor that impressed PSTs' ideas. They think that he pursues the issue by belonging to the city, taking active roles in related processes, and carrying out studies that will benefit the residents. A person's connection with the place he lives in seems effective in the decision-making process related to local issues. When this bond is strengthened with knowledge and experience, a healthier decision-making process will probably be experienced.

At the end of the implementation process that covers interactions with the experts on the issue, it was concluded that PSTs' awareness of the issue was formed; they could consider the issue from different dimensions, and their argumentative skills were improved. More specifically, according to the PSTs, construction of an NPP in Sinop is expected to have positive effects on the economy, meet energy needs, provide development in the field of nuclear medicine, provide job opportunities, be environmentally friendly, and be a political power. Opposing views, on the other hand, included negative effects on the environment, warming of seawater, and foreign dependency in terms of economy. PSTs said that although they had not been inclined to this issue before as an outsider, they want to take action about this issue, transfer the information they learned to their environment, and do more research on it. They also pointed out that they see themselves as more knowledgeable and equipped and that they can put forward many arguments on the issue.

4.3 Transfer to the Global Context

Today, countries like China, India, Russia, and the United States represent half the world's population. For these countries, energy needs are an important issue with the forecast of an increasing population in the future, and it is thought that fossil fuels will be limited in meeting their future demands. Together with ever-increasing energy demand, economic prosperity, and energy security also lead developing countries like

Pakistan, Türkiye, and Bangladesh to use nuclear energy as an alternative to fossil fuels. The developed countries of Asia, such as Japan, South Korea, and Taiwan, have active nuclear reactors for power generation. On the other hand, China is the leading country in the use of nuclear energy with its 39 nuclear reactors and is planning more than 17 new NPPs in Asia. Nuclear reactors are planned and under construction in India, Pakistan, Bangladesh, Türkiye, and the United Arab Emirates (UAE). In addition, a few more countries from Asia, Africa and South America (i.e., Vietnam, Indonesia, Thailand, Philippines, Saudi Arabia, Uzbekistan, Egypt, Jordan, Kenya, Nigeria, Bolivia) are in the process of planning nuclear reactors to provide nuclear energy (Alam et al., 2019). It is foreseen that 164 nuclear reactors will be built by 2030, and there are 317 nuclear reactors in the nuclear energy programs of countries. In the case of the realization of these projects, it is expected that there will be a total of 990 NPPs in the world in 2030 (Çelik et al., 2015).

Planning an NPP construction is undoubtedly an element that affects the residents of the region. Japan, for example, provides 25% of its electricity from nuclear energy. However, the nuclear accidents in the country inevitably made the society's perspective on nuclear power negative (Yıldırım & Örnek, 2007). After the Fukushima accident occurred in 2011, Japanese society opposed the NPP. The nuclear industry has taken many steps and attempts to prevent the opposing tendency, but society's concern continued. With this accident, the public became more sensitive in terms of health and environmental problems (Ramana, 2011). A similar situation has occurred in China. After the Fukushima nuclear accident, public opinion became more serious about nuclear energy, especially due to the uncertainty it created for the global community. The government announced that they were suspending the approval of new nuclear energy projects. Some countries like Japan, Germany, and Switzerland, on the other hand, have started to slow down or even stop the construction of NPPs. Such attempts and accidents caused societal controversy (Sun & Zhu, 2014). In a study investigating opinions about an NPP in Korea (Jho et al., 2014), it is seen that the issue of nuclear energy attracted the attention of the entire public regarding sustainability, but its consequences were discussed. Negative attitudes included fear of leakage of radioactive materials, risk of accidents, risk to health and the environment, and lack of confidence that it is not resistant to natural disasters such as earthquakes. On the other hand, they have also adopted a positive approach for reasons such as preventing excessive use of natural resources, reducing gas emissions, solving economic problems, and finding it appropriate to meet the increasing energy needs (Jho et al., 2014).

Society plays an active role in the processes related to SSIs, especially in the local ones. Studies are done with students, PSTs, teachers, and the public since they are important components in terms of the widespread impact of the issues. In this direction, planning an expert-supported teaching program to develop the arguments of individuals who have a say on local issues was thought to be important. It is thought that the inclusion of teachers and teacher candidates, who appeal to large audiences and have a high impact, in such training improves their understanding of these issues and their arguments and, therefore, their professional development. Local issues can be transferred to the classroom environment within appropriate teaching methods determined in the curriculum. The contributions of the experts in the field provide valuable experiences for students. Providing the acquisitions by experts in many aspects, such as knowledge, experience, resources, and research findings on these issues, contributes to the development of argumentation skills in terms of explaining the issue with arguments in the decision-making process of individuals on local issues. Nuclear has been discussed in Türkiye and in the world, and it will continue to be discussed as long as the need for energy takes place in the public. For this reason, expert support can be applied as a teaching method in the transfer of SSI-based content of curricula in Türkiye and the world. Considering that scientists who carry out scientific studies and are experts in the field have a decisive impact on society, contacts should be made between scientists and societies in order to establish healthy *science communications*. With these communications, local issues can be negotiated, and scientists can have a chance to see public responses and make scientific explanations. Providing scientific communication with experts can deliver information on local issues to large masses. Awareness of local issues can be raised by enabling individuals of all ages to establish a link between science and society and science and technology. In addition, gathering experts and PSTs can contribute to introducing new, current, and original studies and expand the argument network with the support of scientific content, data, and study findings. Similarly, conferences can be given to individuals by experts of differing views. In this way, it can be ensured that they have multiple perspectives and can use the knowledge and experience they have gained while evaluating local issues.

4.4 Conclusion

A place-based SSI instruction was given to a group of PSTs in this study. The local SSI in the study was the construction of an NPP in Sinop. The issue is controversial in the city because while some groups in the region oppose the establishment of the plant by creating an anti-nuclear platform and holding demonstrations regularly, some others support the initiatives regarding the construction of the plant. Being the residents of the city, PSTs were chosen for participation, and they were brought together with the experts on this issue through online meetings moderated by the researchers of this study. Experts provided students with their knowledge, experience, and perspectives about the issue. This scientific communication between the experts and the PSTs revealed valuable contributions to the PSTs about the local SSI in terms of knowledge, awareness, beliefs, and behaviors. It is seen in this study that although having difficulty making a decision by not having enough information on this issue before the implementation, all PSTs held a position after the process and developed several supporting and opposing arguments. They also developed better arguments by evaluating the issue from different dimensions, including environment, economy, science, politics, and technology. PSTs' willingness to take action about the issue and enthusiasm towards transferring the knowledge they learned to their environment were valuable outputs of the place-based SSI instruction they experienced.

References

Akyüz, E. (2017). Advantages and disadvantages of nuclear energy in Turkey: Public perception. *Eurasian Journal of Environmental Research, 1*(1), 1–11.

Alam, F., Sarkar, R., & Chowdhury, H. (2019). Nuclear power plants in emerging economies and human resource development: A review. *Energy Procedia, 160*, 3–10.

Ateş, H., & Saraçoğlu, M. (2013). Fen bilgisi öğretmen adaylarının gözünden nükleer enerji. *Ahi Evran Üniversitesi Kırşehir Eğitim Fakültesi Dergisi, 14*(3), 175–193.

Barnes, D. (1992). *From communication to curriculum* (2nd ed.). Portsmout Boynton/Cook Publishers.

Bricker, L., & Bell, P. (2009). Conceptualizations of argumentation from science studies and the learning sciences and their implications for the practices of science education. *Science Education, 92*, 473–498.

Çapkınoğlu, E., & Yılmaz, S. (2018). Yedinci sınıf öğrencilerinin yerel sosyobilimsel konulardaki argümanlarında kullandıkları very bileşeninin incelenmesi [Examining the data component used by seventh grade students in arguments related to local socioscientific issues]. *Eğitim ve Bilim, 43*(196), 125–149.

Çelik, İ., Çeker, A., & Belge, R. (2015). Nükleer enerji: Türkiye ve dünya ölçeğinde bir değerlendirme [Nuclear energy: An evaluation of Turkey and World Scale]. *Yeni Fikir Dergisi, 7*(15), 55–68.

Christenson, N. (2015). *Socioscientific argumentation: Aspects of content and structure* (Doctoral dissertation). Retrieved from http://urn.kb.se/resolve?urn=urn:nbn:se:kau:diva-35869

Demircioğlu, T., & Uçar, S. (2014). Investigation of written arguments about Akkuyu nuclear power plant. *Elementary Education Online, 13*(4), 1373–1386.

Driver, R., Asoko, H., Leach, J., Mortimer, E., & Scott, P. (1994). Constructing scientific knowledge in the classroom. *Educational Researcher, 23*(7), 5–12.

Duschl, R., Ellenbogan, K., & Erduran, S. (1999, March). *Promoting argumentation in middle school science classrooms: A Project SEPIA evaluation.* A paper presented at the Annual Meeting of the National Association for Research in Science Teaching, Boston.

Eral, M. (2015). *Nükleer güç santralleri ve ülkemiz*. Retrieved April 2023, from https://www.kmo.org.tr/resimler/ekler/1423037b0f99b51_ek.pdf

Eş, H., & Varol, V. (2019). Fen bilgisi öğretmenliği ve ilahiyat öğrencilerinin nükleer santral sosyo-bilimsel konusuyla ilgili informal argümanları [The informal argumentation of theology and science education students about the socio-scientific issue: Nuclear power plant]. *Mersin University Journal of the Faculty of Education, 15*(2), 437–454.

Es, H., & Yenilmez Türkoglu, A. (2021). Science and non-science majors' mental models of nuclear power: Does the program of study matter? *Acta Didactica Napocensia, 14*(2), 231–243.

Eş, H., Işık Mercan, S., & Ayas, C. (2016). Türkiye için yeni bir sosyo-bilimsel tartışma: nükleer ile yaşam [A new socio-scientific issue for Turkey: Life with nuclear]. *Turkish Journal of Education, 5*(2), 47–59. https://doi.org/10.19128/turje.92919

Evren Yapıcıoğlu, A., & Kaptan, F. (2018). Sosyobilimsel durum temelli öğretim yaklaşımının argümantasyon becerilerinin gelişimine katkısı: Bir karma yöntem araştırması. *Ondokuz Mayıs Üniversitesi Eğitim Fakültesi Dergisi, 37*(1), 39–61. https://doi.org/10.7822/omuefd.278052

Herman, B. C. (2018). Students' environmental NOS views, compassion, intent, and action: Impact of place-based socioscientific issues instruction. *Journal of Research in Science Teaching, 55*(4), 600–638.

Iordanou, K., & Constantinou, C. P. (2014). Developing preservice teachers' evidence-based argumentation skills on socio-scientific issues. *Learning and Instruction, 34*, 42–57.

Jho, H., Yoon, H. G., & Kim, M. (2014). The relationship of science knowledge, attitude and decision making on socio-scientific issues: The case study of students' debates on a nuclear power plant in Korea. *Science & Education, 23*, 1131–1151.

Kılınç, A., Boyes, E., & Stanisstreet, M. (2013). Exploring students' ideas about risks and benefits of nuclear power using risk perception theories. *Journal of Science Education and Technology, 22*, 252–266.

Kuhn, D. (2010). Teaching and learning science as argument. *Science Education, 94*(5), 810–824. https://doi.org/10.1002/sce.20395

Kuhn, D., Zillmer, N., Crowell, A., & Zavala, J. (2013). Developing norms of argumentation: Metacognitive, epistemological, and social dimensions of developing argumentative competence. *Cognition and Instruction, 31*(4), 456–496. https://doi.org/10.1080/07370008.2013.830618

Lee, Y. C., & Grace, M. (2010). Students' reasoning processes in making decisions about an authentic, local socio-scientific issue: Bat conservation. *Journal of Biological Education, 44*(4), 156–165. https://doi.org/10.1080/00219266.2010.9656216

Lin, S. S., & Mintzes, J. J. (2010). Learning argumentation skills through instruction in socioscientific issues: The effect of ability level. *International Journal of Science and Mathematics Education, 8*, 993–1017.

Mathew, M. D. (2022). Nuclear energy: A pathway towards mitigation of global warming. *Progress in Nuclear Energy, 143*. https://doi.org/10.1016/j.pnucene.2021.104080

McClennen, N. (2016, May 9). *Place-based education: Communities as learning environments*. Getting Smart. https://www.gettingsmart.com/2016/05/09/place-based-education-communities-as-learning-environments/

MENR. (2023). Retrieved April 2023, from https://enerji.gov.tr/bilgi-merkezi-enerji-elektrik

Ozturk, N., & Bozkurt Altan, E. (2019). Examining science teachers' decisions about nuclear power plants from the perspective of normative decision theory. *Journal of Education in Science, Environment and Health (JESEH), 5*(2), 192–208. https://doi.org/10.21891/jeseh.581739

Ramana, M. V. (2011). Nuclear power and the public. *Bulletin of the Atomic Scientists, 67*(4), 43–51.

Sadler, T. (2004). Informal reasoning regarding socioscientific issues: A critical review of research. *Journal of Research in Science Teaching, 41*, 513–536. https://doi.org/10.1002/tea.20009

Sun, C., & Zhu, X. (2014). Evaluating the public perceptions of nuclear power in China: Evidence from a contingent valuation survey. *Energy Policy, 69*, 397–405.

Tippett, C. (2009). Argumentation: The language of science. *Journal of Elementary Science Education, 21*(1), 17–25.

Türköz, G., & Öztürk, N. (2020). Fen bilgisi öğretmen adaylarının bazı sosyo-bilimsel konularla ilgili kararlarının çok boyutlu bakış açısı ile incelenmesi [Examination of pre-service science teachers' decisions about some socio-scientific issues with a multidimensional point of view]. *Cumhuriyet Uluslararası Eğitim Dergisi, 9*(1), 175–197.

Walton, D. N. (1989). Dialogue theory for critical thinking. *Argumentation, 3*, 169–184.

Yenilmez Türkoğlu, A., & Öztürk, N. (2019). Sosyo-bilimsel konulara ilişkin fen bilgisi öğretmen adaylarının zihinsel modelleri [Pre-service science teachers' mental models of socio-scientific issues]. *Başkent Üniversitesi Eğitim Fakültesi Dergisi, 6*(1), 127–137.

Yıldırım, M., & Örnek, İ. (2007). Enerjide son seçim: Nükleer enerji [Ultimate choice for energy: The nuclear energy]. *Gaziantep Üniversitesi Sosyal Bilimler Dergisi, 6*(1), 32–44.

Yılmaz, E. (2015). Güvenlik ve ekonomik boyutuyla nükleer enerji tartışmaları: Akkuyu nükleer santrali örneği [Nuclear energy discussions in terms of economic and security dimension: Reactions against Akkuyu nuclear plant]. *Cumhuriyet Üniversitesi Fen-Edebiyat Fakültesi Sosyal Bilimler Dergisi, 39*(1), 227–245.

Zeidler, D. L., Herman, B. C., & Sadler, T. D. (2019). New directions in socioscientific issues research. *Disciplinary and Interdisciplinary Science Education Research, 1*, 11. https://doi.org/10.1186/s43031-019-0008-7

Zohar, A. (2008). Science teacher education and professional development in argumentation. In S. Erduran & M. P. Jiménez-Aleixandre (Eds.), *Argumentation in science education: Perspectives from classroom-based research* (pp. 245–268). Springer. https://doi.org/10.1007/978-1-4020-6670-2_12

Chapter 5
Preservice Primary School Teachers' Experience of Place-Based Education: Porsuk River

Ahmet Oğuz Akçay and Mehmet Arif Bozan

5.1 Introduction

The slogan "There is no other world to live in!" is gaining popularity. This slogan emphasizes issues such as increasing global warming, environmental pollution, and unplanned urbanization, implying that significant natural problems await humanity soon. Water quantity and quality are two of these issues (UNESCO, 2009). In terms of water quantity, it is estimated that at least two billion people are unable to access safe drinking water due to a lack of rainfall and rising temperatures (Gleick, 2019). Water pollution is another SSI. Among the important factors contributing to the deterioration of water quality are industrial organizations, incorrect agricultural activities, and urbanization (Dwivedi, 2017). Water pollution causes economic and social harm by disrupting the services provided by the ecosystems formed by rivers to the people living in that region (Best, 2019; Palmer & Ruhi, 2019). As a result, it is extremely necessary to protect water, which is a critical resource for the region.

Rivers are an important water source and provide numerous benefits to the region in which they are located, including agricultural, tourism, economic, and other benefits. However, human activity pollutes rivers, which can be found in city centers as well as other water resources. Unplanned urbanization is the first human cause. Precipitation and temperatures above normal cause events such as floods and

A. O. Akçay (✉)
Department of Primary School Education, Faculty of Education, Eskişehir Osmangazi University, Eskişehir, Türkiye
e-mail: aoakcay@ogu.edu.tr

M. A. Bozan
Department of Primary School Education, Faculty of Education, Istanbul Aydın University, İstanbul, Türkiye
e-mail: mehmetbozan@aydin.edu.tr

B. Namdar, E. Karahan (eds.), *Socioscientific Issues Focused Teacher Education*, Contemporary Trends and Issues in Science Education 60,
https://doi.org/10.1007/978-3-031-55233-5_5

overflows as a result of city population growth and urbanization. Another issue is household waste. The untreated discharge of wastewater from settlements into rivers pollutes the environment. When household waste is dumped into rivers without being treated, it causes chemical, physical, and biological pollution. Agricultural activities are also significant contributors to river pollution. Chemical fertilizers used in agricultural areas are transported to rivers as a result of precipitation, etc., or the mixing of vegetable solid wastes from agricultural areas into rivers pollutes the water. The introduction of chemical fertilizers into rivers disrupts a river's ecological balance. The wastewater from industrial organizations is one of the major causes of river pollution. The negative effects of industrial waste on the environment are far greater than those of other causes. It is well known that the liquid waste of these organizations pollutes not only the water but also the soil and vegetation, causing environmental destruction.

The pollution of rivers in city centers has negative consequences. Pollution in a river will have a negative economic impact on the region where it is located due to factors such as a decrease in aquaculture products and a decrease in tourism. River pollution can also cause heart disease, chronic respiratory diseases, developmental and nervous system disorders, and immune system disorders in the region's residents. On the other hand, it disrupts the river's ecological environment and kills the living creatures that inhabit it. Given the importance of rivers to a region, it is critical to raise awareness among the people who live there through education.

5.2 SSI Context: Pollution in Porsuk River

Various situations affecting people and nature are arising today as a result of developing science and technology, transportation, population growth, and so on (Brandt et al., 2013; Gibbs, 2017; Moore et al., 2018). Individuals are expected to achieve global citizenship awareness to solve SSI that concern society (Brandt et al., 2013; Gibbs, 2017; Moore et al., 2018). Individuals can achieve the desired profile, according to this viewpoint, by incorporating SSI into education.

SSI are contentious and complex social issues that can be addressed scientifically and have multiple solutions (Sadler, 2004). SSI allow students to apply theoretical knowledge in the real world, defend and discuss their ideas, and solve problems in their daily lives (Carson & Dawson, 2016; Eastwood et al., 2011). As a result, including SSI in educational content will be an important tool for students to be sensitive to societal issues and develop skills such as problem-solving. In this context, ensuring preservice teachers' professional development is critical to integrating SSI into education effectively.

In terms of educational quality, teacher education has been important throughout history. In this context, to teach the importance of SSI to students in a qualified manner, teachers must first gain this awareness before beginning their professional careers. According to the literature, teachers have misconceptions and limited perspectives about SSI (Türkmen et al., 2017; Yilmaz Türkoğlu & Öztürk, 2019) and are insufficient in preparing lesson plans for teaching SSI (Evagorou, & Puig Mauriz, 2017). As a result, it is critical to raise preservice teachers' awareness of SSI, as well as their competency in teaching it.

The purpose of this study was to raise preservice primary school teachers' awareness of river pollution in the region where they live, as well as design solutions to the identified pollution. When the literature was examined, it was discovered that preservice teachers lack knowledge about water pollution and are unaware of its significance (Kariper, 2014). With a total length of 448 km, 225 km of the Porsuk River (Ministry of Environment and Urbanization, 2021) passes through the region where the preservice teachers participating in the study were educated. The Porsuk River is a valuable resource for Eskişehir in terms of both tourism and agriculture. This river provides drinking water and is used on a daily basis (Bakış et al., 2011). Despite its importance to the city, pollution from human sources is increasing (Öztürk, 2007; Bakış et al., 2011; Köse et al., 2016). In the Ministry of Environment and Urbanization's report (2021), the annual average nitrate value measured, especially in agricultural areas of the Porsuk River, is around 9. The World Health Organization has stated that the nitrate value in water should not be above 10. Considering that the Porsuk River is also used for drinking water, this situation is risky. The high nitrate values in the Porsuk River in agricultural areas confirm the statement that excessive fertilization used in agricultural areas also causes nitrates. In parallel with this situation, it was revealed that the water was clean at the source of the Porsuk River, but when the quality of the water in the region outside of the province was checked, it was found to be in the class of polluted or highly polluted water due to industrial and domestic waste (Köse et al., 2016). Drought is another factor threatening the Porsuk River. Sarş (2017) examined past low flows in the Porsuk River basin and determined that the basin experienced a hydrological drought in recent years. As a result, it is critical to use such an important resource for the city sustainably.

According to Gruenewald (2003), place-based education has four basic characteristics, and this study was structured around these four characteristics. The first distinguishing feature is the uniqueness of the chosen location. The Porsuk River was chosen as the study location in this context. The Porsuk River runs through the heart of Eskişehir and is vital to the city's agricultural and tourist industries. The second distinguishing feature is that the place-based learning process is multidisciplinary in nature. In the research for a solution to the Porsuk River pollution, solutions were tried using information from science, technology, and engineering. For example, it is a matter of engineering how deep the Pyramid filter should be placed in the Porsuk River. The third feature is the implementation of location-based activities with real-world experiences. Pre-service teachers could better observe pollution and seek solutions in this context by visiting the Porsuk River. They went to the area and took photos of the pollution (see Fig. 5.1). The fourth feature is the structure, which can make pre-service teachers believe place-based activities benefit society. As a result of the activity, the pre-service teachers identified Porsuk's pollution and explained why resolving this issue is critical for the city.

Fig. 5.1 Pollution caused by human beings

5.3 Implementation Procedure

The experiences of preservice primary school teachers in identifying the causes of environmental pollution affecting the Porsuk River in the environment where they were educated were examined in this section. In this context, the goal was to raise preservice primary school teachers' awareness of SSI in their region. The study included 48 preservice primary school teachers studying at a state university in Eskişehir. The preservice primary school teachers were divided into 14 groups of three to four students. Porsuk River pollution is caused by rapid population growth, unplanned urbanization, and industrialization. The preservice teachers were expected to increase their awareness of this problem and design a solution for it in this chapter. The study was completed in 7 weeks in this context. Below are the details of the 7-week training provided to the preservice primary school teachers.

Week 1 The preservice teachers were asked to research environmental pollution in Eskisehir. In this context, the preservice teachers investigated environmental issues affecting Eskisehir and determined that they should concentrate on air, soil, and water pollution. This aimed for the students to become aware of their environment and understand what is negatively affecting it. The preservice teachers were asked to use the university library, local press, people living in Eskisehir and around the Porsuk River, and academic studies for research to determine the environmental pollution affecting Eskisehir. In other words, they were asked to search from different sources. In this context, they had the opportunity to examine the pollution of Eskisehir in a multidimensional way. Primarily, they visited the Porsuk River and conducted an observation of the contamination. The preservice teachers' photographs captured during the observation are displayed in Figs. 5.2 and 5.3.

Week 2 The next stage required the preservice teachers to investigate environmental pollution in the Porsuk River. In this context, they concentrated on two major causes of pollution in the Porsuk River: industrialization (dumping factory wastes into the river) and human-induced pollution (agricultural spraying, littering, and so on). The negative consequences of Porsuk River pollution, the endangerment of living life as a result of pollution, the danger of decreasing the flow rate of the river due to unconscious agricultural irrigation, and the negative impact on human health by spreading foul odors in the environment due to garbage that is thrown into the river all came to the forefront.

Fig. 5.2 Observing river pollution

Fig. 5.3 Different liquids mixing in a river

Week 3 In the third week, students who found the causes of pollution of the Porsuk River were asked to find solutions to the river pollution. The groups formed in this context began to develop alternative solutions to the river's pollution. The aim of this task was for them to realize that the solutions to pollution are very simple and doable.

Week 4 In the classroom, the preservice teachers presented possible solutions to river pollution to the other group members. They discussed the various solutions in class and chose the best ones based on suggestions and recommendations from their peers. For example, some groups mentioned the need for a net in the Porsuk River for human waste. However, some group members mentioned that the nets would negatively affect the fish and other creatures living in the Porsuk River and that the fish and creatures caught in these nets would disappear. They were asked to create a prototype based on the solution they selected. It was determined that most groups concentrated on measures to reduce pollution and clean up waste in the Porsuk River.

Weeks 5–6 The groups created prototypes based on the solution proposals they identified the previous week and presented them to their classmates. The prototypes' shortcomings and strengths were discussed. For example, the group that designed a net to be stretched across the badger Porsuk River to prevent human waste, while other groups mentioned that no attention was paid to living life. In other words, although the net is dangerous for fish and some other aquatic life, there is a possibility that aquatic creatures can get caught in the net and die. A pyramid filter system has been proposed for this group. Another group designed a robot that moves on the water's surface. However, it was suggested that the size of this robot should be suitable for the Porsuk River. Since Porsuk River is also used for tourism and tourist trips are organized on it with gondolas, it is proposed to design a robot that does not affect tourism and is large enough to clean human waste on the surface of the Porsuk River.

Week 7 During the final week of the training, the preservice primary school teachers completed a general evaluation. They shared their experiences related to the causes and solutions of environmental pollution in the Porsuk River.

5.4 Outcomes

In this study, it was determined how the preservice teachers' experiences were realized during the 7-week practice period in Porsuk River. Pollution problems caused by population growth, unplanned urbanization, and industrialization were discussed, and solutions were developed by the preservice teachers. Under this heading, the reasons for river pollution and how the preservice teachers identified these reasons, as well as their suggestions for pollution solutions and the projects they designed to address pollution, were presented.

5.4.1 Sources of Pollution

To begin, the preservice teachers sought to identify and reveal the factors that contribute to river pollution, as well as develop solutions and projects to reduce pollution. They, for example, emphasized the river's long structure and stated that wastewater from industrial organizations in different cities before arriving in the region is one of the crucial factors in pollution by mixing with the river.

> In the discharge waters of various industrial establishments in Kütahya Province, the pH, BOD, COD, AKM, Pb, Cd, oil and grease, and total phosphorus were analyzed between 2003 and 2004. According to the findings, the heavy metal levels in the samples exceeded the Water Pollution Control Regulation's limit values.

Furthermore, in order to better understand environmental pollution, some of the preservice teachers interviewed a person living in that region to learn more about the problems caused by industrial organizations.

> A citizen recalling childhood memories mentioned that the river flowed in the color of the dye used that day in the nearby fabric factories.

Similarly, examples of industrial establishments near the stream were provided to demonstrate the pollution caused by the river. It was stated, for example, that a "porcelain industry, thermal power plant, sugar factory, and industrial organization" were established near the river and poured their wastewater into it. According to reports, this is one of the most serious problems with river pollution.

Furthermore, some of the preservice teachers approached the factors causing river pollution from a different angle. They claimed that invasive fish also contributed to the deterioration of the Porsuk River's natural structure. They stated that invasive fish would pose a significant threat by eating the local fish, reducing biodiversity, and allowing diseases and parasites to enter the habitats. As evidence, they presented an interview with an academic working in the biology department at a local university.

> One of our provinces, Eskişehir, has been severely infiltrated by invasive fish. In Türkiye, invasive fish account for 15%–17% of total freshwater fish, but in Eskişehir, this rate rises to 40%.

Another source of pollution in the Porsuk River is human-generated solid waste. People's waste thrown on the river's surface not only pollutes the river but also disrupts water flow due to waste accumulation.

> Human waste is another factor. Unintentionally thrown waste of all kinds get caught in the flow of the water, causing pollution and blockages in places.

According to some of the preservice teachers, air, water, and soil pollution are all caused by each other. Air pollution caused by human causes reduces the amount or regularity of precipitation, increasing the amount of mud in the river and causing pollution and bad odor.

> Exhaust fumes mix into the air, influencing rainfall. Because of this imbalance, irregular rainfall increases the amount of mud in the stream, resulting in pollution and odor.

Similarly, it was stated that because many people work in the organized industrial zone around the river, vehicle traffic is heavy in that area, negatively impacting air pollution and, thus, the river.

> There are 65 thousand people working in the Organized Industrial Zone, and the ring road cannot handle this vehicle traffic load at the start of the working day and at the end of the working day. The air is heavily polluted as a result of this vehicle density.

In addition, the river is used for irrigation in agriculture and other areas, and the residues of the water used are transferred back to the stream along with the chemicals used in agriculture, which is one of the major causes of pollution.

> The Porsuk River contains important irrigation areas. The rich nitrogen and phosphorus compounds in irrigation water cause pollution in the Porsuk River and Porsuk Dam.

5.4.2 Suggestions for Solutions

Following the identification of pollution causes, the preservice teachers presented general solution proposals. To begin, they made some recommendations for reducing waste generated by humans and industrial organizations, which all of the groups identified as the primary cause of river pollution. According to the preservice teachers, some European countries use treatment devices to clean wastewater from industrial organizations before transferring it to groundwater.

> In developed Western countries such as Germany and France, treated wastewater is cleaned and transferred to groundwater.

The preservice teachers stated that there should be sanctions against polluting industrial organizations and that the public should also react to polluting industrial organizations.

> These polluting factories must be avoided. If financial investments are required...When factories do this, the public should take appropriate actions and issue warnings.

Furthermore, industrial and urban planning should be designed to protect the environment.

> Industrialization and urbanization are required. Natural environmental awareness must be maintained in tandem with industrialization and urbanization. Industry must pay for the pollution it has caused.

Human factors, which are another source of pollution in the river, were also mentioned. In this context, they proposed raising awareness among those living in that region and ensuring public participation in the environmental cleaning or pollution prevention work.

> Awareness-raising activities on environmental pollution should be carried out, and public participation in these activities should be realized.

All stakeholders in the river area should be made aware of the situation. It was stated that "awareness-raising activities should be carried out in schools" for this purpose. On the other hand, it was suggested that farmers who use tea as an irrigation tool in agricultural activities should be given information about products that are appropriate for their region, which will help them use less water.

> The appropriate agricultural method should be chosen based on the type of agriculture and soil condition, and these areas should be monitored on a regular basis by the municipality. Simultaneously, if adequate and less irrigation is performed, the amount of water that must be recycled or transmitted to the basin will be reduced.

Concerning river pollution caused by agricultural practices, they proposed that agricultural areas surrounding the stream install treatment systems and that after using the water, these treatment waters be reused in agricultural areas without mixing with the stream.

> In order to address agricultural pollution in Porsuk Stream, treated water should not be discharged back into the stream. Treatment systems can be installed in agricultural areas, and the process can be completed without dumping into the river.

5.4.3 *Project Designs*

The groups devised projects to address the pollution of the Porsuk River. The preservice teachers preferred to use technological tools to clean up pollution both on the surface and deep within the water. For example, a submerged device was created to clean the Porsuk River (Fig. 5.4). The goal of this device is to clean up the Porsuk River's deep pollution. The device operates via a remote control and separates the

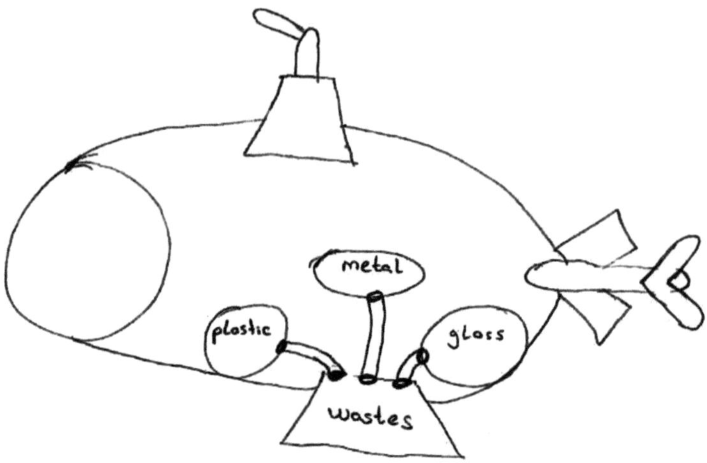

Fig. 5.4 Underwater robot

waste it collects. When the device reaches the surface, it is intended to transfer the waste collected in it to recycling.

Similarly, the preservice teachers preferred to design robots that move on the water's surface and planned to leave these robots at various points along the Porsuk River to move. Its goal is to collect plastic garbage in the Porsuk River alongside the robot. In addition, each robot has its own waste collection area. When the robot's reservoir is full, it will transport the garbage it has gathered to these collection points. The staff will sort the garbage in the area. Toys will be made from recycled plastics. Finally, it was planned to visit primary schools with robots and present recycled toys to the children in order to raise environmental awareness among primary school students.

Another method preferred by the preservice teachers for river cleaning was the collection of solid wastes with various materials to be disposed of in the river (Fig. 5.5). In this context, a pyramid-shaped water waste catcher system, which serves as a filter to clean plastic waste and protect aquatic habitats, was proposed. Water flows between the pyramids, waste is collected in the gaps at the bottom, and the system is kept from becoming clogged. Waste is collected at regular intervals and recycled in recycling facilities.

Similarly, garbage cans were designed to collect trash from the river by vacuuming, eliminating the need for human intervention. The floating trash bins collect one and a half kilograms (around one pound) of waste on the surface of the water on average every day, and their placement in specific areas of the Porsuk River will allow for improved water quality and increased sustainability (Fig. 5.6).

Some of the preservice teachers who wanted to create a more environmentally friendly product planned to collect solid waste on the river's surface with a net that would not harm the river's fish and to separate the collected solid wastes for recycling (Fig. 5.7). Materials that could not be recycled were to be used as fuel in cement plants.

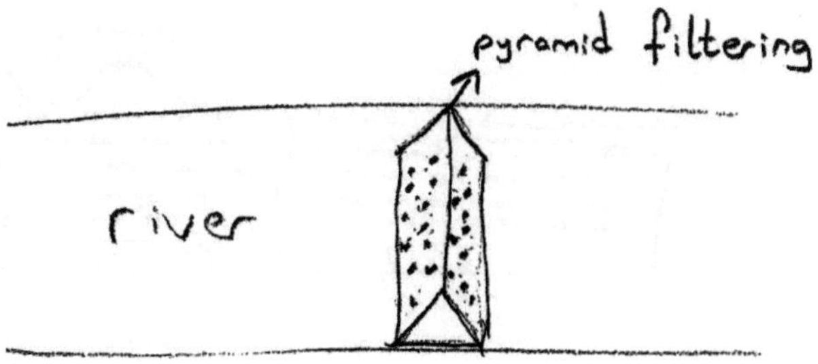

Fig. 5.5 Solid collection vehicles

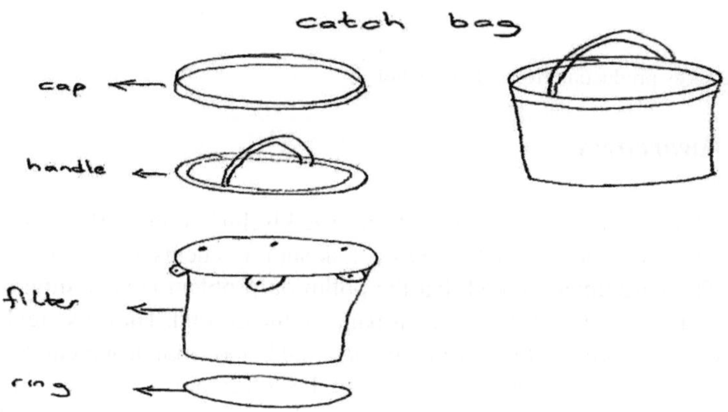

Fig. 5.6 Floating garbage bins

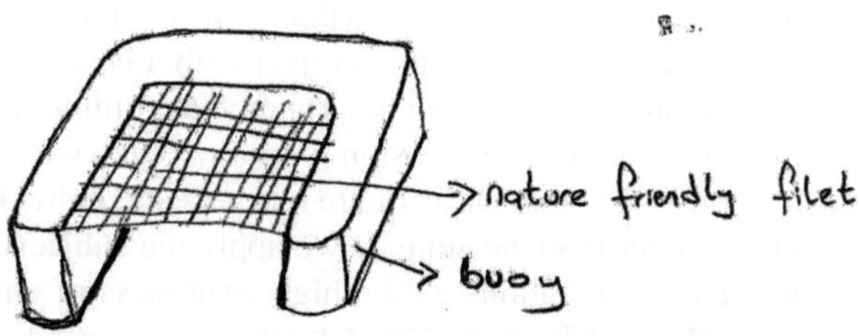

Fig. 5.7 Nature-friendly filet

The placement of "microorganisms that break down detergent active ingredients" in the Porsuk River to clean the detergent active ingredients that come with wastewater was one of the projects designed to solve pollution naturally. As a solution to this problem, it was created to educate detergent manufacturers about the use of biodegradable active ingredients.

Finally, a project to prevent pollution caused by industrial organizations was developed (Fig. 5.8). According to reports, the sugar factory in the area generates spirit as a byproduct. The solution proposed in this context was that the spirit is composed of ethyl alcohol and is renewable, so converting the spirit can meet the factory's energy needs.

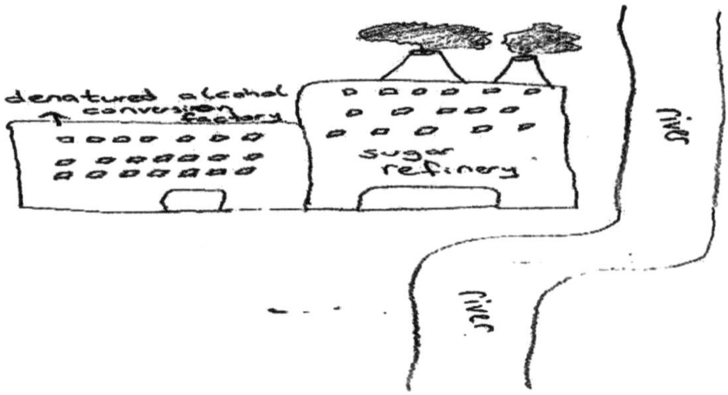

Fig. 5.8 Energy production from ethyl alcohol

5.4.4 *Awareness*

When we look at the causes of pollution in Eskişehir's Porsuk River, we see that there are several, including factory waste, Eskişehir residents, and the river's use for tourism. The participants stated that the pollution problem is very simple and that pollution is more than just throwing garbage on the ground. They also realized that, as prospective teachers, they were unconcerned about environmental pollution, acted irresponsibly, and made no effort to find a solution.

Aside from human-caused pollution, the participants realized how much factory waste pollutes the river and agreed that the most important thing is to keep it clean. In summary, the participants stated that most environmental pollution prevention projects to date have focused on cleaning up the polluted environment and that by replacing such projects with ones that focus on not polluting our environment at all, and when we can carry out this project with a healthy system, we can minimize or even eliminate the problem of environmental pollution. In this context, environmental pollution, a topic of discussion for many years, has recognized the importance of raising conscious students as future teachers and individuals who are aware of keeping clean rather than cleaning. If we apply the subjects we explain to our children, we will have generations with a high awareness of environmental pollution if we become role models.

5.5 Transfer to the Global Context

Water is a source of life for all living things, not just humans. Water resources are dwindling due to both human and natural causes. As a result, over two billion people face water scarcity (United Nations, 2018). Because resources are limited, the quality of the remaining water resources is critical. The vast majority of industrial wastewater is estimated to be transferred to regional water resources in a way that disrupts the order of people and ecosystems (United Nations, 2021). The wastewater from industrial organizations is also discharged into the water resources of the region in which they are located, depleting already scarce water resources. As a result, improving both the quantity and quality of water resources is critical.

This section focused on the pollution of the Porsuk River in Eskişehir. In particular, although we focused on the Porsuk River in this study, the preservice teachers mentioned the rivers in their hometowns and the pollution affecting them during the training. In many countries, rivers are passing through the middle of cities (e.g., Paris – Seine River, Rome – Tiber River, etc.). Hence, the activities carried out in this study can be used in many countries. In this context, education is an essential tool. Along with education, it is critical to raise children's awareness of water resources in their communities and teach them how to protect them. For children to gain this awareness formally, teacher education programs should be designed in such a way that preservice teachers' awareness of environmental issues, such as water resources in their region, is included. It is critical for preservice teachers to become aware of water resource issues in their communities and become pedagogically competent in how to incorporate these issues into their lessons. Teachers should help students become more aware of the water resources in their area, as well as the scientific knowledge gained in the process and the argumentation and decision-making processes involved in making water resources more efficient (Nielsen, 2009).

5.6 Conclusion

Rivers are impacted by pollution caused by increased urbanization and industrialization. Water resources are polluted when people throw waste into the environment and when domestic and industrial wastewater from industry and urbanization is discharged untreated into rivers. As a result, people's access to clean water is becoming increasingly difficult.

The purpose of this study is to raise future teachers' awareness of river pollution. In this context, prospective teachers will be able to inform their future students about this issue if they are aware of it. This can be achieved through place-based education, where students learn about their immediate surroundings and develop a connection to their environment. By incorporating environmental awareness education into the early stages of schooling, we can lay a strong foundation for students' understanding of environmental issues. In this regard, it is essential to enhance the environmental awareness of prospective teachers. By equipping them with knowledge and sensitivity towards environmental matters, they will be better prepared to impart this awareness to their students once they become teachers. Ultimately, by instilling a sense of environmental responsibility in both educators and students, we can foster a generation that is mindful of and actively contributes to the well-being of their surroundings. Raising environmental awareness from a young age will result in a more livable environment.

References

Bakış, R., Koyuncu, H., Özkan, A., Banar, M., Yılmaz, G., & Yörükoğulları, E. (2011). An investigation of the surface and groundwater pollution level in the Porsuk Basın. *Anadolu University Journal of Science and Technology Applied Sciences and Engineering, 12*(2), 75–89.

Best, J. (2019). Anthropogenic stresses on the world's big rivers. *Nature Geoscience, 12*, 7–21. https://doi.org/10.1038/s41561-018-0262-x

Brandt, P., Ernst, A., Gralla, F., Luederitz, C., Lang, D. J., Newig, J., & Von Wehrden, H. (2013). A review of transdisciplinary research in sustainability science. *Ecological Economics, 92*, 1–15. https://doi.org/10.1016/j.ecolecon.2013.04.008

Carson, K., & Dawson, V. (2016). A teacher professional development model for teaching socioscientific issues. *Teaching Science, 62*(1), 28–35.

Dwivedi, A. K. (2017). Research in water pollution: A review. *International Research Journal of Natural and Applied Sciences, 4*, 118–142.

Eastwood, J. L., Schlegel, W. M., & Kristin, L. C. (2011). Effects of an interdisciplinary program on students' reasoning with socioscientific issues and perceptions of their learning experiences. In T. D. Sadler (Ed.), *Socioscientific issues in the classroom: Teaching, learning and research, contemporary trends, and issues in science education* (pp. 89–126). Springer.

Evagorou, M., & Puig Mauriz, B. (2017). Engaging elementary school pre-service teachers in modeling a socioscientific issue as a way to help them appreciate the social aspects of science. *International Journal of Education in Mathematics, Science and Technology, 5*(2), 113–123. https://doi.org/10.18404/ijemst.99074

Gibbs, P. (2017). *Transdisciplinary higher education: A theoretical basis revealed in practice*. Springer.

Gleick, P. H. (2019). Water as a weapon and casualty of conflict: Freshwater and international humanitarian law, water resources management. *Springer Nature, 33*, 1737–1751.

Gruenewald, D. A. (2003). Foundations of place: A multidisciplinary framework for place conscious education. *American Educational Research Journal, 40*(3), 619–954.

Kariper, İ. A. (2014). Çevre eğitiminde su ve su kirliliğinin önemi [The importance of water and water pollution in environmental education]. *Journal of European Education, 4*(1), 19–22.

Köse, E., Çiçek, A., Emiroğlu, Ö., Tokatlı, C., Uğurluoğlu, A., Başkurt, S., Aksu, S., & Uylaş, M. (2016). Water quality assessment of Porsuk stream basin. *Biological Diversity and Conservation, 9*(3), 119–126.

Ministry of Environment and Urbanization. (2021). *Eskişehir ili çevre durum raporu* [Eskisehir province environmental status report]. https://cedbi.cekulvakfi.org.tr/CEDBI_pdf_files/Eskisehir_Il_Cevre_Durum_Raporu.pdf

Moore, M., Martinson, M. L., Nurius, P. S., & Kemp, S. P. (2018). Transdisciplinarity in research: Perspectives of early career faculty. *Research on Social Work Practice, 28*(3), 254–264. https://doi.org/10.1177/1049731517708033

Nielsen, J. A. (2009). Structuring students' critical discussions through processes of decision making on socio-scientific controversies. *Revista de Estudos Universitários, 35*(2), 139–165.

Öztürk, R. (2007). *Porsuk Creek environmental problems and watershed management suggestions in solutions to these problems*. Unpublished master's thesis. Çukurova University.

Palmer, M., & Ruhi, A. (2019). Linkages between flow regime, biota, and ecosystem processes: Implications for river restoration. *Science, 365*(6459), 1249–1263. https://doi.org/10.1126/science.aaw2087

Sadler, T. (2004). Informal reasoning regarding SSI: A critical review of research. *Journal of Research in Science, 41*(5), 513–536.

Sarış, F. (2017). Low flow analysis in Porsuk creek basin. *İstanbul Üniversitesi Edebiyat Fakültesi Coğrafya Dergisi, 33*, 72–82.

Türkmen, H., Pekmez, E., & Saglam, M. (2017). Pre-service science teachers' thoughts about socioscientific issues. *Ege Eğitim Dergisi, 18*(2), 448–475. https://doi.org/10.12984/egeefd.295597

UNESCO. (2009). *The United Nations world water development report 3: Water in a changing world*. Berghahn Books.

United Nations. (2018). *Progress on transboundary water cooperation 2018 global baseline for sdg indicator 6.5.2*. https://www.unwater.org/publications/progress-transboundary-water-cooperation-global-baseline-sdg-indicator-652

United Nations (2021). *World water development report 2021: Valuing water*. https://www.unesco.org/reports/wwdr/2021/en

Yılmaz Türkoğlu, A., & Öztürk, N. (2019). Pre-service science teachers' mental models of socio-scientific issues. *Başkent University Journal of Education, 6*(1), 127–137.

Chapter 6
Preservice Teachers' Enactment of Character and Values Through a Place-Based Drama Activity

Ayşegül Oğuz Namdar, Bahadır Namdar, and Fulden Guler Nalbantoglu

6.1 Introduction

6.1.1 Character and Values for Twenty-First-Century Citizenry

Rapid developments in science and technology necessitate the evaluation of ethical and moral consequences in individual, local, and global dimensions in human life and within the ecosystem (Lee et al., 2012). Especially in the twenty-first century, science education should not only focus on students' acquisition of scientific knowledge and basic skills but also the development of character, moral values, and virtues (Choi et al., 2011; Lee et al., 2013). In other words, students should develop their character and values as global citizens.

Choi et al. (2011) state that character and values should be included within the re-conceptualization of science literacy. They define character and values as follows: "Belief system and preferences to act responsibly with respect to human life and compassion for other human beings throughout the globe. This belief system serves as a driving force for individuals to act responsibly as citizens of technologically and scientifically advanced 21st-century society" (p. 681). According to this definition, character and values are examined in three sub-dimensions. The ecological worldview is the belief that individuals are connected to the ecosystem.

A. Oğuz Namdar (✉)
Department of Primary Education, Faculty of Education, Recep Tayyip Erdoğan University, Rize, Türkiye
e-mail: aysegul.oguz@erdogan.edu.tr

B. Namdar · F. Guler Nalbantoglu
Department of Mathematics and Science Education, Faculty of Education, Ege University, Izmir, Türkiye
e-mail: bahadir.namdar@ege.edu.tr; fulden.guler@ege.edu.tr

B. Namdar, E. Karahan (eds.), *Socioscientific Issues Focused Teacher Education*, Contemporary Trends and Issues in Science Education 60,
https://doi.org/10.1007/978-3-031-55233-5_6

Individuals with this worldview recognize the interdependence of environmental components and the necessity for all beings to exist sustainably. Social and moral compassion encompasses empathy and respect shown towards other living beings. Individuals who possess this value are inclined to associate with empathy and moral reasoning processes concerning those affected by the issue. Having social and moral compassion involves having moral and ethical sensitivity, understanding the different positions and interests of stakeholders involved, and empathizing with those affected by the situation. Socioscientific responsibility, on the other hand, is defined as individuals feeling responsible for problem-solving and their willingness to take action in this regard.

Lee et al. (2012) emphasize the importance of equipping students in today's schools with sensitivity toward their world, ethical sensitivity towards global SSI, and the skills to use and evaluate their own character and values with empathetic feelings. The acquisition of such character and values is seen as crucial for students to make evidence-based informed decisions (Zeidler & Sadler, 2011). Within the context of SSI, character and values are seen as essential reference points for individuals to act responsibly and make informed decisions regarding these issues (Choi et al., 2011; Herman et al., 2018).

Character and value development can be facilitated by SSI teaching approaches, according to findings from a growing number of empirical research conducted in recent years. However, only a few studies have been found regarding this topic with preservice teachers. For instance, Lee et al. (2012) examined the impact of an SSI teaching program on the character and values of 18 Korean preservice teachers, focusing on nuclear energy, climate change, and embryonic stem cells. The study found that the preservice teachers utilized all three dimensions of character and values in their reasoning, but they were not able to consistently apply them and had limited evaluation of global perspectives. In their research, Uzel (2020) examined the changes in the character and values of preservice biology teachers after the SSI curriculum. The research results showed that the character and values of preservice teachers developed in all sub-dimensions. In another study, Öztürk (2019) determined the characters and values of 201 preservice teachers as global citizens by survey method. Accordingly, preservice teachers got the highest score in the category of empathic concerns, while the lowest score was in the category of feeling responsible.

6.1.2 *Drama in Socioscientific Issue-Based Education*

Drama has long been one of the favored teaching methods used by science educators. With drama, teachers engage in analogical reasoning and modeling in different science topics (Harrison & Treaugust, 2000); teach domain-specific knowledge related to various science subjects (Oğuz Namdar et al., 2018); create active discussion and argumentation environments (Archila, 2017; Verhoeff, 2017); and positively influence students' affective characteristics towards the lesson (Ong et al., 2020).

It is also observed that drama techniques are frequently utilized in the teaching processes of SSI topics. According to Odegaard (2003), improvisational activities like role-writing or reenacting current events can be used in the classroom to incorporate drama and encourage students to make decisions about the societal implications of science. Particularly, through role-playing processes, students can explore situations that foster empathy and identification, thus indicating an increase in their willingness to take action on the examined topics. Oğuz Namdar and Namdar (2021) have stated that drama techniques can support socioscientific argumentation processes. Empirical studies also support this notion. Sharing arguments and presenting different perspectives are facilitated through role-playing in SSI contexts with conflicting situations. For example, role-playing processes regarding vaccination have been found to enhance students' argument quality about SSI (Maniatakou et al., 2020). In another study, drama processes regarding genetic testing were integrated with university students. The research findings revealed that drama supports socioscientific argumentation processes. Archila et al. (2023a) engaged university students in socioscientific argumentation processes through drama on genetic testing. The research indicated that drama creates an environment for students to generate arguments and counter-arguments, thus improving argumentation quality. Likewise, Archila et al. (2023b) included drama processes regarding genetically modified foods, and the research found that drama improved argumentation skills.

Drama not only enables students to engage in argumentation processes but also allows them to consider different perspectives, multiple dimensions of an event, and alternative explanations. In this process, students utilize not only their knowledge but also their character and values, becoming informed decision-makers and ready to take action as twenty-first-century citizens. According to Odegaard (2023), drama and role-playing activities positively influence affective aspects and facilitate students' involvement in effective decision-making processes. Additionally, semi-structured drama activities require students to utilize knowledge from different disciplines and incorporate their values and ethical concerns.

6.1.3 Place-Based SSI and "Character and Values"

Place-based learning is an educational approach that brings students together with the local environment and community through active and realistic learning experiences (Powers, 2004). van Eijck and Roth (2010) describes it as "a teaching–learning process that centers on what is considered local—usually students' own place, that is, their immediate schoolyard, neighborhood, town, or community (p. 187). In place-based learning, the "place" is situation-specific. The place near school, home, and community creates a learning environment in which students can transfer their knowledge to daily-life issues in a practical and meaningful way (Gosselin et al., 2016). Through place-based learning, students have the opportunity to conduct comprehensive investigations into local issues and enhance their subject matter knowledge related to these issues (Demarest, 2014). Additionally, place-based learning helps students develop a deeper understanding of their surroundings beyond the boundaries of the school and encourages them to take an active role in addressing local issues (Gruenewald, 2003; Sobel, 2004).

On the other hand, a sense of place is defined as "the meanings of and the attachments to a place held by a person or group" (Semken, 2005, p. 149). Lim and Barton (2006) suggest that developing an attachment to a specific place and comprehending its significance can enhance engagement and motivation in science learning. The sense of place shapes how individuals perceive and understand nature, providing students with authentic settings and hands-on experiences (Gosselin et al., 2016). Interacting with the environment and people helps students to broaden their understanding of the sense of place (Langran & Dewitt, 2020).

6.1.4 Drama, Character and Values, & Place-Based Education

Drama not only enables students to engage in argumentation processes but also allows them to consider different perspectives, multiple dimensions of an event, and alternative explanations. In this process, students utilize not only their knowledge but also their character and values, becoming informed decision-makers and ready to take action as twenty-first-century citizens. According to Odegaard (2023), drama and role-playing activities positively influence affective aspects and facilitate students' involvement in effective decision-making processes. Additionally, semi-structured drama activities require students to utilize knowledge from different disciplines and incorporate their values and ethical concerns.

Drama can utilize several mechanisms to enhance character and values for informed decision-making as twenty-first-century democratic citizenship. Firstly, drama can engage students in perspective-taking (Mages, 2018). Drama activities provide opportunities for students to immerse themselves in different roles, contemplate challenging situations, and experience diverse viewpoints. This process, in turn, develops students' empathy toward others' perspectives. Developing perspective-taking skills is crucial for fostering empathy in resolving SSI (Newton & Zeidler, 2020). Perspective-taking involves individuals considering how others feel and perceive issues based on different values in order to justify moral actions (Kahn & Zeidler, 2016). Since drama activities involve perspective-taking, they can promote the consideration of values and, consequently, the cultivation of those values. Second, because SSI have ethical dimensions, and moral and ethical reasoning is an unavoidable part of socioscientific decision-making (Zeidler & Nichols, 2009), drama activities provide a space for people to explore the ethical dimensions of such issues. Role-playing, for example, can engage students in emotive, intuitive, and ethical reasoning, which helps develop students' character and values concerning various SSI. Third, drama encourages personal reflection on issues, and personal stories, for example, can be used to consider moral values (Gervais, 2006). Students are encouraged to consider their own values, experiences, and biases regarding SSI through drama. This critical reflection and evaluation process can then help students develop their own values and reasoning processes. Fourth, drama activities can help students develop socioscientific responsibility and motivate them to take action to solve SSI. Drama activities can motivate students to apply their character and values to real-world problems and thus become agents of change. They actively participate in problem-solving discussions and in addressing complex SSI.

According to researchers, through place-based experiences, students form affective connections with their surroundings and see themselves as an integral part of the environment, which encourages pro-environmental behavior (Herman et al., 2018, 2020). Drama activities in place-based education can allow students to immerse themselves in the environment and have firsthand reflective and authentic experiences with a real-life problem, fostering responsible and empathetic individuals for positive change by employing character and values (Avşar Erümit et al., 2023). The importance of incorporating place-based pedagogies to foster the effect of drama activities in developing character and values for informed decision-making about SSI is emphasized in this study.

6.2 Implementation

6.2.1 Context

Izmir is a coastal city located in the Aegean Region of Türkiye, with a population of over four million. The districts of Bornova and Bayraklı in Izmir are among the most cosmopolitan areas of the city, hosting numerous high-rise buildings, modern residences, shopping centers, hospitals, two universities, and many industrial establishments. However, being an area that still experiences significant migration, it is situated on a plain with abundant water resources. Particularly since the 1950s, it has been used for urbanization. These districts were affected by a major earthquake in 2020, with its epicenter in Samos, Greece. As a result of the earthquake, four buildings were utterly destroyed, while hundreds of buildings were damaged. Despite the fertile lands and dense population in these districts, high-rise construction continues, and the recent earthquake disaster has once again brought the issue of urbanization in this region to the forefront, sparking numerous debates.

6.2.2 Participants

Within the scope of this study, a total of nine preservice teachers participated in the activity voluntarily. Two of the preservice teachers are male, and seven are female. Preservice teachers were enrolled in the final year of the undergraduate science teaching program of a large research university located in Bornova, Izmir. Previous to the current study, the preservice teachers completed practical activities on drama and SSI within the scope of courses such as environmental education and science teaching and gained theoretical knowledge about these methods. However, none of them had experience with place-based learning. In-person participation in these extracurricular activities took place during the spring semester of 2022–2023.

6.2.3 Procedures

The major earthquake tragedy in Türkiye caused the spring semester of 2022–2023 to be held in a hybrid format in universities. Because of this, most preservice science teachers completed their coursework virtually during this time. After contacting and informing the preservice teachers who lived in Izmir about the study, the activities were conducted at a time that worked best for them.

The activities took a full day. The study was completed in four stages. In the first stage, students were given the scenario: Tall Buildings or Agriculture? (Box 6.1). Students were asked about their arguments about whether the area should be opened to agricultural activities by reducing the construction activities in the Bornova Plain.

Box 6.1 Scenario Given to Preservice Teachers

Bornova is one of Izmir's most populous districts, with a population of around 450,000 people (http://www.izmir.gov.tr/ilcelerimiz). Bornova's population has grown in recent years due to an increase in the number of business and residential centers, as well as extensive migration. The district currently has two universities, Ege University and Yaşar University, the region's largest hospital, Ege University Hospital, as well as numerous hospitals, public and private schools, industrial establishments, entertainment and shopping centers, and luxury residential projects. It is now one of the most popular districts due to its easy transportation and proximity to the city centers of Izmir and Manisa, as well as various industrial establishments.

Yeşilova Mound, Izmir's first historical settlement, is located in Bornova (Karadaş, 2014). When the structural characteristics of Bornova are examined, however, it is discovered that it is only 50 meters above sea level. However, the plain formed by alluvial accumulation has seen human settlement over its 8500-year history (Karadaş & Öner, 2021). However, the recent construction of high-rise buildings and the resulting population increase clearly demonstrates that the plain, which is also rich in water resources, has been used for construction, particularly since the 1950s (Doygun & Erdem, 2012).

However, in recent years, these development activities have resulted in some disagreements among farmers, urban and regional planners, city center workers, and environmentalists. Opening the Bornova Plain to settlement provides many benefits to the people who live there, such as socialization, easy

(continued)

Box 6.1 (continued)
access to health and service sectors, and convenient transportation. However, it has a negative impact on farmers' production activities and causes environmental damage. Furthermore, Izmir is an earthquake city; on October 30, 2020, an earthquake with a magnitude of 6.9 struck the Aegean Sea, causing many buildings in the Bornova district to collapse or be severely damaged. This earthquake also brought the issue of construction in the Bornova district back to the forefront.

Do you think that construction activities in the Bornova Plain should be reduced and the region should be opened to agricultural activities? Write your argument in detail with your justifications.

In the second stage, the preservice teachers visited the area affected by the 2020 earthquake disaster in Bornova, Izmir, where many buildings were destroyed or damaged, resulting in the deaths of many people. The students were divided into groups of three and instructed to conduct informal interviews with local residents and business owners in the region to learn about the feelings, experiences, and opinions of those who had been affected by the earthquake. During the interviews, preservice teachers were given the following questions to ask local residents:

- What are your thoughts on opening agricultural lands in Bornova for settlement?
- Do you think these regions should be used for agriculture instead of these buildings?
- What do you think has changed economically, ecologically, and socially in the region since agricultural lands were made available for settlement?
- How happy are you with the new situation?
- Do you believe the settlement should have been relocated to a safer location outside of Bornova following the disaster?
- Determine how people's perceptions of the region's construction changed before and after the earthquake.

During this stage, the preservice teachers took brief notes and recorded local residents' perspectives on the subject. The preservice teachers returned to campus for the third stage. They were given a 30-minute period on campus to fill out Table 6.1 with the positive and negative aspects of the environmental, social, economic, and other impacts of the Bornova Plain settlement. They were then asked to participate in a small group discussion about their groups' initial arguments about the issue as a result of their interviews and information-gathering process.

Drama activities began in the fourth stage. Drama activities were carried out during the stages of preparation, enactment, and evaluation. The preservice teachers were divided into pairs during the preparation phase for the Adaptation-Trust-Empathy activities. One of the students placed a hand on the other's shoulder and gave various commands to move the blindfolded student safely through the space.

Table 6.1 Argument construction paper

Do you think construction activities in the Bornova Plain should be reduced and the region open to agricultural activities?			
Yes No			
Interview data supporting your decision	Interview data opposing your decision	Research data supporting your decision	Research data opposing your decision
Your argument:			

Students were thus exposed to the values of empathy, trust, and responsibility. The role-playing technique was used during the improvisation phase. The students were divided into two groups and assigned to the roles listed below. The students were told that they were guests of a TV program about the Izmir Earthquake and that they would be speakers at a panel about reducing the construction activities in the Bornova plain and opening the region to agricultural activities, and they were told to choose one of the following roles or prepare their own roles (Table 6.2).

While the students played their roles, the researchers assumed the role of a TV program host and conducted the discussion. After the role-play, the evaluation phase started. In the evaluation phase, which was carried out in the form of a group discussion, their opinions on the construction of agricultural lands in the earthquake zone, their evaluations of the day, and their opinions on how the place-based learning approach affected their values on the subject were taken.

6.2.4 Data Collection and Analysis

Within the scope of the study, data were collected through students' worksheets, their responses to the scenario before and after the activities, and a reflection questionnaire consisting of five open-ended questions filled in after the activity. These questions are: What did you learn about construction in the earthquake zone at the end of the place-based learning activity? What effect did the drama activities have on your learning about the positive and negative aspects of this topic? What do you think your individual responsibility is regarding the topic? How do you think the

Table 6.2 Roles given to preservice teachers and questions asked to be considered

A: A 32-year-old entrepreneur. Her family owns land in Bornova. Aylin wants to continue agricultural activities in the region because her family has been involved in agriculture for centuries
B: A 45-year-old contractor. He has built many residential projects in the region. He wants the few remaining agricultural lands in the area to be zoned and insists on earthquake-resistant buildings
C: A 35-year-old environmental engineer. He is concerned about the environmental impact of development in the area and is an activist for the protection of natural habitats in the Bornova region
D: A 40-year-old economist. He believes that the construction in the region and the employment created by large shopping mall projects will benefit the economy of the region and the country
Please also evaluate the following situations during the role-playing process
Ecological worldview
Reduction of sustainable areas and degradation of ecological habitats as a result of the problem of building on fertile agricultural land
Use of renewable energy sources in the construction and use of buildings, including green areas in building areas
Socioscientific responsibility
Possible ethical and social consequences of building in an earthquake zone
Socially positive results of the construction in the region
Exactly what to do to solve the problem individually
Social and moral compassion
Consider how different people might be affected by this
Empathically consider what these people might think and feel in this situation

activities and decision-making processes you are involved in regarding the topic affect your values on this topic? (Ecological worldview, Social and Moral Compassion, Socioscientific Responsibility) How would you use place-based learning approaches in your future teaching practices? Within the scope of this book chapter, the participants' responses to the scenario and reflection questions were analyzed. Content analysis was used to analyze the data (Yıldırım & Simsek, 2006).

6.2.5 Findings

Participants' responses to the scenario before and after the drama activity regarding the components of character and values were summarized in Box 6.1. According to the table, there was a modest change in participants' ecological worldview after the drama activities. They frequently mentioned the interrelationship between humans and the environment by the end of the activity. Additionally, the findings indicated that none of the participants referred to sustainable development prior to the activity, but two of them started emphasizing its importance after the drama activity.

Table 6.3 The frequency of the responses before and after the drama activity concerning character and values

	Conceptual component	Example Item	f(n) (Before)	f(n) (After)
Ecological worldview	Interconnectedness	Construction should continue, but many living organisms will be affected by the destruction of green areas	4	5
	Sustainable development	By using renewable energy sources in Bornova Plain, construction activities can be carried out without damaging the green areas too much. People can produce their own food. Moreover, earthquake-resistant, and environmentally friendly buildings can be built	–	2
Social and moral compassion	Moral and ethical sensitivity	Due to urbanization that has been going on for a very long time, the people living there claim that it is impossible to cultivate agricultural lands. They argue that people would never want to leave the area and that the same social and economic conditions cannot be provided	3	3
	Perspective-taking	When we conducted on-site inspections and spoke with the local shopkeepers in the earthquake-stricken area, they expressed that it would not be suitable for them to move elsewhere because they believed they would not have the same opportunities there. I agree with them on this matter	–	3
	Empathic concerns	There are too many buildings in Bornova, and if the construction is stopped and the existing areas are opened for agricultural activities, there will be limited space left for people. Therefore, it is important to demolish risky buildings and relocate those individuals while opening agricultural land. I think it is crucial for people to feel safe during this process.	–	1

(continued)

Table 6.3 (continued)

	Conceptual component	Example Item	f(*n*) (Before)	f(*n*) (After)
Socioscientific accountability	Feeling of responsibility	I can say that my sense of responsibility has developed in terms of being conscious and increasing awareness regarding the opening of the region for agricultural activities	–	2
	Willingness to act	The region should not be opened for agricultural activities. However, if buildings pose a threat to the environment, I will participate in various projects to protect the environment	–	1

Regarding the second component of character and values, participants' social and moral compassion improved after the drama activity, as observed in Table 6.3. The findings specifically showed that participants enhanced their perspective-taking skills through the drama activity. They began considering alternative points of view regarding the settlement on the agricultural land. However, their empathic concern showed only a slight change at the end.

Concerning the socioscientific accountability component, the findings suggested that the drama activities had a positive impact on participants' feelings of responsibility and willingness to take action, as none of the participants had discussed these aspects before the activity. The example items for each conceptual component after attending the activity were displayed in Table 6.3. In summary, engaging in short-term drama activities triggered the development of participants' character and values. Therefore, it is recommended to conduct long-term activities to support participants in internalizing the experience.

The analysis of reflection papers demonstrated that participants' character and values improved at the end of place-based drama activities. Firstly, the participants' responses were gathered into two categories concerning their understanding of settlement in earthquake-prone areas: content knowledge and the importance of on-site experiences. The findings revealed that the majority of participants expanded their understanding of the importance of constructing solid, enduring, and resilient buildings by considering the geological characteristics of earthquake-prone areas at the end of the activity. Additionally, many participants noted that their understanding was enhanced in terms of comprehending different perspectives on the settlement in agricultural areas through their on-site experience in the place-based learning activity. For instance, P5 explained,

> I learned from the perspectives of people who have experienced this situation (suffered from the earthquake) in the area, real estate agents, and shop owners that the buildings in earthquake-prone areas are highly vulnerable to earthquakes. They are often constructed as multi-story structures, and valuable, fertile agricultural land is being destroyed.

It could be concluded from the findings that place-based learning supports participants in connecting content knowledge and real-life situations and promotes authentic context for learning.

Secondly, participants' responses can be categorized into two with respect to the effect of drama activities on learning positive and negative aspects of the issue: perspective taking and understanding ecological impact. Participants began considering alternative viewpoints and gaining a better understanding of opposing ideas after participating in the drama activity. For example, P8 said, "With the drama activity, I actually thought from a different perspective that I had never thought about before." Similarly, the other participant expressed that he changed his mind after being involved in the drama activity and developed an understanding of opposite ideas. He said, "I believed that agricultural land should be created before the drama activity; however, now I think the settlement should continue in this area (P6)." Additionally, participants talked about how their understanding improved concerning the ecological impact of settlement in agricultural areas. For example, P1 stated, "Through this activity, my knowledge has increased about the positive effects of maintaining agricultural land, such as preserving natural balance and ensuring a healthy future for our next generations."

Thirdly, two overarching categories were found in participants' responses regarding their individual responsibilities on the issue: conducting awareness-raising activities and willingness to act. Before attending the activity, participants did not mention their individual responsibilities related to settlement in the agricultural area. However, their feeling of responsibility increased as they engaged in place-based drama activities. For instance, P1 said, *"I should organize consciousness-raising activities about the settlement in the agricultural areas for my students in the future."* Moreover, P5 indicated, *"I can participate in activities that oppose the destruction of productive lands and environmental damage."* In this way, it could be stated that the participants started to realize how they could individually contribute to the issue after attending place-based drama activity in this study.

Fourth, participating in the place-based drama activities increased participants' character and values in several ways. Their responses centered on moral and ethical sensitivity, increased feelings of responsibility, perspective taking, and emphatic concerns about components of character and values in the reflection papers. Most participants explained that they felt more responsible towards the environment at the end of the activity. Similarly, most participants mentioned that they improved their understanding of the needs, concerns, and diverse viewpoints of the local community that suffered from the earthquake. Lastly, some participants wrote that their empathy towards these people developed after completing the activity. For example, one participant wrote,

> I was affected by the psychology of individuals who had experienced earthquakes. I felt deeply saddened when I saw the eyes of an uncle whose house had been damaged in the earthquake welling up with tears, and I made an effort to empathize with him (P6).

Lastly, regarding integrating place-based learning into teaching, two categories emerged as follows: informal learning environments and SSI. For instance, one of the participants indicated,

> I would like to take my students to relevant places for units like Human-Environment (in the Turkish science curriculum). As we realized today, it provides a better learning experience. For this reason, I will use environments such as museums, zoos, planetariums, and science centers, which are suitable for the place-based learning approach (P1).

Another participant explained, "*While teaching* SSI, *I can take students to the actual locations where issues occur. In this way, students can gather different perspectives from various individuals. This makes the learning more effective and permanent" (P5).* As a result, participants' understanding of place-based learning started to enhance, and they started to touch upon how they might utilize it in their teaching at the end of the activity.

6.3 Transfer to the Global Context

Urbanization of agricultural land is one of the pressing issues of Türkiye. Due to changes in population density, industrialization, and tourism, fertile agricultural areas are increasingly being opened to construction (Duran et al., 2012; Tanrıvermiş, 2003). This is an even more pressing issue, as there is a high number of immigrants to Türkiye, especially in urban areas and big cities such as Istanbul, Ankara, and Izmir in recent years. However, this issue is neither new nor only a local problem for Türkiye. Across the world, many countries have been facing similar problems for over a couple of decades (Pandey & Seto, 2015; Beckers et al., 2020; Hou et al., 2021). Therefore, it is both a national and a global SSI that demands the engagement of individuals in a discussion of the issues both at the local and global scales, as the importance of agricultural fertile land use has been better understood due to high food prices across the globe. Hence, individuals should have character and values as global citizens to tackle the issue for a better and sustainable world.

The place-based drama activities in this specific study are transferable to other contexts. In this study, we chose a locally relevant and highly contemporary topic due to recent earthquakes in Türkiye. However, drama activities have been proven to offer a space for expressing value judgments, therefore, enacting character and values in the contexts of energy issues (Namdar and Oğuz Namdar, 2021), climate change (Namdar and Oğuz Namdar 2022), and evolution (Duveen & Solomon, 1994). However, none of these studies benefited from "place" and place-based instruction. Therefore, drama activities can be blended with place-based SSI so that the preservice teachers can use their character and values for informed decision-making.

6.4 Conclusion

In this study, we merged drama and place-based activities to foster preservice teachers' character and values about the urbanization of agricultural land use. There has been increasing interest in using drama in SSI-based teaching and learning. Yet, empirical evidence and concrete tools for implementation are still new and scarce. Therefore, this study could offer an example of the confluence of two areas of research: place-based education and drama. Although the initial results are promising, one should be aware of the fact that the activities are only implemented over one day with a limited number of participants. We still saw some of the immediate reflections from the implementation. However, future studies should investigate the effect of the activities on the enactment of character and values and the development over an extended period. Moreover, we only used unstructured role plays in this study to engage students in drama activities. Future studies might employ more structured activities.

References

Archila, P. (2017). Using drama to promote argumentation in science education. The case of "Should've". *Science & Education, 26*(3–4), 345–375. https://doi.org/10.1007/s11191-017-9901-7

Archila, P. A., Restrepo, S., Truscott de Mejía, A. M., & Bloch, N. I. (2023a). Drama as a powerful tool to enrich socio-scientific argumentation. *International Journal of Science and Mathematics Education, 21*(5), 1661–1683. https://doi.org/10.1007/s10763-022-10320-3

Archila, P. A., Truscott de Mejía, A. M., & Restrepo, S. (2023b). Using drama to enrich students' argumentation about genetically modified foods. *Science & Education, 32*(3), 635–668.

Avşar Erümit, B., Namdar, B., & Oğuz Namdar, A. (2023). *Using place-based SSI instruction that utilize role-playing to promote preservice teachers' socioscientific accountability and NOS.* Paper presented at the 96th NARST International Conference, Chicago, IL, USA.

Beckers, V., Poelmans, L., Van Rompaey, A., & Dendoncker, N. (2020). The impact of urbanization on agricultural dynamics: A case study in Belgium. *Journal of Land Use Science, 15*(5), 626–643. https://doi.org/10.1080/1747423X.2020.1769211

Choi, K., Lee, H., Shin, N., Kim, S., & Krajcik, J. (2011). Re-conceptualization of scientific literacy in South Korea for the 21st century. *Journal of Research in Science Teaching, 48*(6), 670–679. https://doi.org/10.1002/tea.20424

Demarest, A. B. (2014). *Place-based curriculum design: Exceeding standards through local investigations.* Routledge.

Doygun, N., & Erdem, Ü. (2012). Bornova İlçesi'nde alan kullanım potansiyeli ile alan kullanım yapısı arasındaki etkileşimlerin belirlenmesi [Determining interactions between land use potential and land use structure in Bornova District]. *İnönü Üniversitesi Sanat ve Tasarım Dergisi, 2*(5), 141–150.

Duran, C., Gunek, H., & Sandal, E. K. (2012). Effects of urbanization on agricultural lands and river basins: Case study of Mersin (south of Turkey). *Journal of Environmental Biology, 33*(2), 363–371.

Duveen, J., & Solomon, J. (1994). The great evolution trial: Use of role play in the classroom. *Journal of Research in Science Teaching, 31*, 575–582.

Gervais, M. (2006). Exploring moral values with young adolescents through process drama. *International Journal of Education & the Arts, 7*(2), 1–34.

Gosselin, D., Burian, S., Lutz, T., & Maxson, J. (2016). Integrating geoscience into undergraduate education about environment, society, and sustainability using place-based learning: Three examples. *Journal of Environmental Studies and Sciences, 6*, 531–540.

Gruenewald, D. A. (2003). The best of both worlds: A critical pedagogy of place. *Educational Researcher, 32*(4), 3–12.

Harrison, A. G., & Treagust, D. F. (2000). A typology of school science models. *International Journal of Science Education, 22*(9), 1011–1026. https://doi.org/10.1080/095006900416884

Herman, B. C., Sadler, T. D., Zeidler, D. L., & Newton, M. H. (2018). A socioscientific issues approach to environmental education. In G. Reis & J. Scott (Eds.), *International perspectives on the theory and practice of environmental education: A reader. Environmental discourses in science education* (Vol. 3). Springer.

Herman, B. C., Zeidler, D. L., & Newton, M. (2020). Students' emotive reasoning through place-based environmental socioscientific issues. *Research in Science Education, 50*, 2081–2109. https://doi.org/10.1007/s11165-018-9764-1

Hou, D., Meng, F., & Prishchepov, A. V. (2021). How is urbanization shaping agricultural land-use? Unraveling the nexus between farmland abandonment and urbanization in China. *Landscape and Urban Planning, 214*, 104170. https://doi.org/10.1016/j.landurbplan.2021.104170

Kahn, S., & Zeidler, D. (2016). Using our heads and HARTSS*: Developing perspective-taking skills for socioscientific reasoning (*humanities, ARTs, and social sciences). *Journal of Science Teacher Education, 27*(3), 261–281. https://doi.org/10.1007/s10972-016-9458-3

Karadaş, A. (2014). İzmir'in bilinen ilk yerleşmesi Yeşilova Höyüğü'nde jeoarkeoloji ve paleocoğrafya araştırmaları [Paleogeographic and geoarchaeological researches in Yeşilova mound, the oldest known settlement of İzmir]. *Ege Coğrafya Dergisi, 23*(1), 43–55.

Karadaş, A., & Oner, E. (2021). 30 Ekim 2020 Sisam depreminin İzmir-Bayraklı'da yol açtığı hasar üzerinde Bornova Ovasının alüvyal jeomorfolojisinin etkileri [effects of the alluvial geomorphology on the damage of the Sisam earthquake in the Bornova plain]. *Cografya Dergisi, 42*, 139–153. https://doi.org/10.26650/JGEOG2021-872890

Langran, E., & DeWitt, J. (2020). *Navigating place-based learning*. Springer International Publishing.

Lee, H., Chang, H., Choi, K., Kim, S. W., & Zeidler, D. L. (2012). Developing character and values for global citizens: Analysis of pre-service science teachers' moral reasoning on socioscientific issues. *International Journal of Science Education, 34*(6), 925–953. https://doi.org/10.1080/09500693.2011.625505

Lee, H., Yoo, J., Choi, K., Kim, S. W., Krajcik, J., Herman, B. C., & Zeidler, D. L. (2013). Socioscientific issues as a vehicle for promoting character and values for global citizens. *International Journal of Science Education, 35*(12), 2079–2113. https://doi.org/10.1080/09500693.2012.749546

Lim, M., & Barton, A. C. (2006). Science learning and a sense of place in an urban middle school. *Cultural Studies of Science Education, 1*(1), 107–142. https://doi.org/10.1007/s11422-005-9002-9

Mages, W. K. (2018). Does theatre-in-education promote early childhood development?: The effect of drama on language, perspective-taking, and imagination. *Early Childhood Research Quarterly, 45*, 224–237. https://doi.org/10.1016/j.ecresq.2017.12.006

Maniatakou, A., Papassideri, I., & Georgiou, M. (2020). Role-play activities as a framework for developing argumentation skills on biological issues in secondary education. *American Journal of Educational Research, 8*(1), 7–15. https://doi.org/10.12691/education-8-1-2

Namdar, B., & Oğuz Namdar, A. (2021). Fostering students' values through role play about socioscientific issues. *The Physics Teacher, 59*(6), 497–499. https://doi.org/10.1119/5.0019320

Namdar, B., & Oğuz Namdar, A. (2022). Role reversal to facilitate social & moral compassion: A case for climate change as an ethics dilemma. *The American Biology Teacher, 84*(4), 242–245. https://doi.org/10.1525/abt.2022.84.4.242

Newton, M. H., & Zeidler, D. L. (2020). Developing socioscientific perspective taking. *International Journal of Science Education, 42*(8), 1302–1319. https://doi.org/10.1080/09500693.2020.1756515

Ødegaard, M. (2003). Dramatic science. A critical review of drama in science education. *Studies in Science Education, 39*(1), 75–102. https://doi.org/10.1080/03057260308560196

Ødegaard, M. (2023). Using drama in science education and for sustainability issues. In *Learning science through drama: Exploring international perspectives* (pp. 69–86). Springer International Publishing.

Oğuz Namdar, A., & Namdar, B. (2021). Blending creative drama and computer supported collaborative learning for socioscientific argumentation. In W. Powell (Ed.), *Socioscientific issue based instruction for scientific literacy development* (pp. 132–160). IGI Global.

Oğuz Namdar, A., Bülbül, A. N., & Çankal, A. O. (2018). Yaratıcı drama ile canlılar dünyasına yolculuk. *Ahi Evran Üniversitesi Kırşehir Eğitim Fakültesi Dergisi, 19*(2), 1404–1422. https://doi.org/10.29299/kefad.2018.19.02.009

Ong, K. J., Chou, Y. C., Yang, D. Y., & Lin, C. C. (2020). Creative drama in science education: The effects on situational interest, career interest, and science-related attitudes of science majors and non-science majors. *Eurasia Journal of Mathematics, Science and Technology Education, 16*(4). https://doi.org/10.29333/ejmste/115296

Öztürk, N. (2019). Character and values for global citizens: A study with preservice science teachers. *Başkent University Journal of Education, 6*(2), 345–352.

Pandey, B., & Seto, K. C. (2015). Urbanization and agricultural land loss in India: Comparing satellite estimates with census data. *Journal of Environmental Management, 148*, 53–66. https://doi.org/10.1016/j.jenvman.2014.05.014

Powers, A. L. (2004). An evaluation of four place-based education programs. *The Journal of Environmental Education, 35*(4), 17–32. https://doi.org/10.3200/JOEE.35.4.17-32

Semken, S. (2005). Sense of place and place-based introductory geoscience teaching for American Indian and Alaska Native undergraduates. *Journal of Geoscience Education, 53*(2), 149–157. https://doi.org/10.5408/1089-9995-53.2.149

Sobel, D. (2004). *Place-based education: Connecting classrooms and communities*. The Orion Society.

Tanrivermis, H. (2003). Agricultural land use change and sustainable use of land resources in the Mediterranean region of Turkey. *Journal of Arid Environments, 54*(3), 553–564. https://doi.org/10.1006/jare.2002.1078

Uzel, N. (2020). Impact of a SSI program on prospective teachers' character and values for global citizens. *International Journal of Progressive Education, 16*(5), 1–16. https://doi.org/10.29329/ijpe.2020.277.1

van Eijck, M., & Roth, W. M. (2010). Towards a chronotropic theory of "place" in place-based education. *Cultural Studies of Science Education, 5*(4), 869–898. https://doi.org/10.1007/s11422-010-9278-2

Verhoeff, R. P. (2017). The use of drama in socio-scientific inquiry-based learning. In K. Hahl, K. Juuti, J. Lampiselka, A. Uitto, & J. Lavonen (Eds.), *Cognitive and affective aspects in science education research* (pp. 117–126). Springer International Publishing AG. https://doi.org/10.1007/978-3-319-58685-4_9

Yıldırım, A., & Şimşek, H. (2006). *Sosyal bilimlerde nitel araştırma yöntemleri [qualitative research methods in social sciences]* (6th ed.). Seçkin Yayıncılık.

Zeidler, D. L., & Nichols, B. H. (2009). Socioscientific issues: Theory and practice. *Journal of Elementary Science Education, 21*(2), 49–58. https://doi.org/10.1007/BF03173684

Zeidler, D. L., & Sadler, D. L. (2011). An inclusive view of scientific literacy: Core issues and future directions of socioscientific reasoning. In C. Linder, L. Ostman, D. A. Roberts, P. Wickman, G. Erickson, & A. MacKinnon (Eds.), *Promoting scientific literacy: Science education research in trans- action* (pp. 176–192). Routledge/Taylor & Francis Group.

Chapter 7
Developing Preservice Science Teachers' Conceptual Understanding Through Place-Based SSI Instruction: Population Growth in Istanbul

Seda Usta Gezer, Çiğdem Çıngıl Barış, and Filiz Avcı

7.1 Introduction

7.1.1 Conceptual Understanding in the Context of a Place-Based Socioscientific Issue

Concepts can be described as perceptions of objects, events, or phenomena being gathered or defined under a particular perceived label or name (Novak, 2010). Individuals learn through concepts to organize the events and objects they encounter in their minds in a structured and appropriate manner, to use their memories more efficiently, and to remember what they have learned when they need it (İnel Ekici, 2014). Concepts are mental tools that allow individuals to make sense of extensive information by organizing it (Senemoğlu, 2001). During this process, new information is restructured in mind. If that new information does not fit in with the existing knowledge, individuals may modify their knowledge to adapt to new experiences (Bachtold, 2013). The processes of learning and restructuring in the human mind continue throughout one's lifetime (Cunningham & Turgut, 1996). Concepts that enable individuals to perceive their environments and think more healthily undergo changes through experiences, and new experiences gain meaning in the light of previously acquired concepts (Morris, 1996). Novak (2010) stated that concept learning occurs in two ways, namely through concept formation and concept assimilation. Individuals learn concepts based on their characteristics and examples while forming those concepts in their minds. As concepts are assimilated, new concepts

S. Usta Gezer (✉) · Ç. Çıngıl Barış · F. Avcı
Department of Mathematics and Science Education, Hasan Ali Yücel Education Faculty, İstanbul University-Cerrahpaşa, İstanbul, Türkiye
e-mail: sedausta@iuc.edu.tr; ccingil@iuc.edu.tr; filizfen@iuc.edu.tr

B. Namdar, E. Karahan (eds.), *Socioscientific Issues Focused Teacher Education*, Contemporary Trends and Issues in Science Education 60,
https://doi.org/10.1007/978-3-031-55233-5_7

are structured by associating them with existing concepts in the individual's mind. Learning concepts correctly will also affect the correct learning of new information.

Conceptual understanding is defined as the ability to transfer knowledge to different and previously encountered situations (Darmofal et al., 2002). Sinan (2007) described conceptual understanding as deep learning, in which relationships and similarities between concepts can be clearly demonstrated. These concepts can be transferred to new environments when necessary, and they can be used to solve problems encountered in daily life. Conceptual understanding, which is known as the building block of knowledge, and the correct learning of concepts are essential in science education. In teaching SSI related to controversial or conflicting topics, and in cases where students have difficulty grasping concepts, various opportunities should be provided for students to learn these concepts by doing and experiencing them in the context of a particular place. Thus, it is thought that contemporary trends such as the Place-Based Learning (PBL) approach will be increasingly important for science education.

7.1.2 Conceptual Understanding in the Context of Preservice Science Teacher Education

The most significant responsibility in raising conscientious, science-literate individuals rests with science teachers. In this context, university students who will subsequently become science teachers have responsibilities in terms of being knowledgeable about SSI and being prepared to educate future generations as individuals who are sensitive to social events. The high level of conceptual understanding that preservice teachers may acquire in their teaching processes will help both themselves and their future students in perceiving and solving the problems they may encounter (Gödek & Polat, 2017).

Although one of the important goals of science is conceptual understanding, it is known that many students at every grade level struggle to understand many scientific concepts (Gobert & Clement, 1999). The identification and elimination of these difficulties are possible when teachers have gained sufficient knowledge and skills in the field of conceptual understanding (Akgün et al., 2005). The recognition and application of the methods and techniques to be used in this field are crucial for effective conceptual learning processes (Köse et al., 2003). Different learning methods and techniques in which students actively participate can be implemented to ensure the correct understanding of various concepts (Darmofal et al., 2002; Demirel, 2016; Green, 2012; Kong et al., 2023).

PBL applications provide teachers with opportunities for positive professional development, allowing their students to learn more about their community and the environment (Adams et al., 2014; Ebersohn & Ferreira, 2012; Jennings et al., 2005; Stickney, 2020). Dubel and Sobel (2008) explained that place-based teacher education is one of the fundamental steps of place-based education. If teachers understand the PBL approach well, they can effectively provide opportunities for their students' learning experiences. SSI have properties that support students in relating what they learn in school to their own lives with the meaningful questioning and learning of concepts (Dawson, 2015). The PBL approach has the potential to affect conceptual understandings of SSI positively. Thus, it is believed that ensuring conceptual understanding through PBL in educational processes will increase the skills that preservice teachers gain in the course of their own education due to its importance in those educational processes.

7.2 SSI Context: Population Growth as a Socioscientific Issue: The Case of Istanbul

Population is a concept signifying the community of people living in a defined area with precise boundaries at a certain time, such as a continent, country, city, town, or village (Bayat, 2010; Doğan, 2015). Historically speaking, populations have been increasing since the beginning of time. With the emergence of different lifestyles, the growing global population has begun to affect almost every aspect of our lives and has led to changes in living spaces. In particular, population increases have resulted in increased consumption and production, as well as environmental pollution. It is also observed that population growth does not make positive contributions to the environment as long as environmental facilities remain fixed; instead, it has negative effects and makes environmental management more challenging (Bozkurt, 2013).

While the population growth rate in developed countries is increasing by about 0.5–1%, it is increasing at higher rates of about 2–3% in developing countries. Rapid population growth causes economic, social, and environmental problems, leading to inadequate supplies of resources in developing countries and slowing down the development rate of these countries (Çamurcu, 2005; Görmez, 2015; Topal, 2011). Population growth also brings with it urbanization. Population growth has always been a problem in underdeveloped countries alongside urbanization dynamics (Es & Ateş, 2010). According to a report from the United Nations, 55% of the world's population now lives in urban areas, and this rate is expected to reach 70% in 2050. It is predicted that the global population will reach approximately 8.5 billion in 2030, with an additional 1.18 billion added in the following 20 years to reach 9.7 billion in 2050. This situation is expected to increase the trend of urbanization even more (United Nations, 2022).

Türkiye's population is one of the fastest growing in the world, with a population of 85.3 million as of 2022. It is estimated that the population of the country will reach approximately 96 million by 2060 (Kutsal & Polatoğlu, 2023). Like many other countries, population growth in Türkiye brings with it various challenges. Rapid population growth has led to significant urbanization with large-scale migration from rural areas to cities. In Türkiye, the urbanization rate was about 25% in the 1950s, but it reached 77% in 2012. In addition to population growth, migration from rural areas to cities has increased the demand for housing in urban areas. The rapid urbanization of cities has led to problems, including the spread of informal settlements, environmental degradation, and the deterioration of the existing infrastructure (Urban Transformation Research Report, 2014).

Istanbul, Türkiye's most populous city, is a center of attraction due to its historical and natural beauty as well as its status as the country's center of tourism, finance, and culture. Istanbul steadily attracts migrants due to its geopolitical importance. The population of Istanbul, which was about 1 million in the 1950s, increased to 3 million due to rapid migration in the 1970s, indicating uncontrolled population growth. In 1980, the population of the city was 4.741.890, and that number rapidly increased to 7.309.190 in 1990. The significant increase in population continued through 2000 when the population of Istanbul reached about 11 million (Istanbul Metropolitan Municipality, 2001). In the last 20 years, the population growth rate of Istanbul has slowed, and as of 2021, the population of the city was about 16 million (Turkish Statistical Institute, 2022). Istanbul is one of the most striking examples of rapid urbanization due to the fast population growth caused by internal migration. The urbanization elements that have emerged there, mainly due to demographic changes, are also reflected as unplanned or distorted urbanization and environmental problems. Some of the environmental problems caused by population growth include the depletion of water resources, air pollution, and noise pollution. Moreover, the effects of disasters such as earthquakes are felt on a larger scale due to the urbanization caused by population growth.

Humanity is faced with a serious water crisis due to the rapid population growth on a world scale and the resulting unplanned urbanization, as well as the inability to develop efficient water policies. Water holds a significant place in sustaining life within ecosystems due to its role in the hydrological cycle, whereby the water on Earth moves with the help of solar energy. All living organisms on Earth obtain the water they need from this cycle and return it to the same cycle after using it. This cycle ensures that a certain amount of water is available in a given area. However, the amount of water provided by this cycle will not be able to meet the needs of steadily increasing populations. This situation may lead to the depletion of water resources, resulting in a decrease in the amount of water per capita and potential water scarcity in some countries (Yüksek, 2004). According to the Falkenmark water stress index developed by Falkenmark et al. (1989), based on the amount of water per capita, regions with annual per capita water availability below 500 m^3 are classified as experiencing "absolute water scarcity," those with 500–1000 m^3 experience "water scarcity," those with 1000–1700 m^3 experience "water stress" and those with more than 1700 m^3 experience "rare water scarcity." Türkiye, despite being bordered by water on three sides, has become a country that experiences water stress due to the effects of migration, industrialization, and population growth in recent years. According to the Turkish Statistical Institute (TSI), the annual per capita usable water in Türkiye is expected to decrease to 1120 m^3 by 2030 (Türkyılmaz, 2010). Istanbul, which is Türkiye's most populous city and continues to experience population growth, is expected to be among the cities that will feel the effects of that water scarcity the most.

One of the problems brought about by the population growth in Istanbul is air pollution. Air pollution, which is defined as the presence of any chemical, physical, or biological factor that alters the natural composition of the indoor and outdoor atmosphere (World Health Organization, 2022), constitutes the most significant and easily spread type of pollution that makes countries responsible to each other (Sofuoğlu, 2003). Air pollution, which originally arose from natural causes such as volcanic eruptions and forest fires, has reached serious levels today due to human activities. This situation varies according to the level of urbanization and industrialization and the population size of different places (Menteşe, 2017). Istanbul, as Türkiye's most populous city, is among the cities of the world where air pollution is felt intensely due to the presence of industrial centers, heavy traffic, and highly polluting emissions from domestic fuels. According to a previous study, 31.476 deaths occurred in Türkiye in 2019 due to air pollution, and it was determined that those deaths could have been prevented if the air pollution limit set by the World Health Organization had not been surpassed (Cooperation Project for Environment, Climate and Health, 2022). Considering that Istanbul's population will continue increasing day by day, it is inevitable that the negative effects of air pollution will continue to be seen.

In Istanbul, where various industrial and other human activities take place, noise pollution has also reached levels that cause discomfort for both people and the broader environment, like other types of pollution. Noise pollution is also referred to as sound pollution (Morillas et al., 2002; Zannin et al., 2002). Noise is generally defined as an unwanted, unpleasant, and intolerable sound (Watts, 1995). Although there are many reasons for noise pollution, its main causes are rapid population growth, industrialization, and the consequent rapid and chaotic urbanization. The main sources of noise pollution are traffic noise, outdoor activities such as sports areas and markets, indoor and outdoor workplaces and industrial noise, construction noise, and commercial noise (Özyonar & Peker, 2008). Given that Istanbul has all these types of noise, the negative effects of noise pollution are inevitable as the population continues to increase.

Another concept closely related to population growth is urban renewal. One important factor that makes urban renewal necessary is the possibility of earthquakes. The world's bitter experiences with earthquakes have revealed that urban renewal is urgently needed. The identification and elimination of unhealthy building stock before an earthquake is likely to minimize the loss of life and property in a possible earthquake. The fact that a large part of Türkiye is located in a first-degree earthquake zone makes disaster-sensitive urban renewal a necessity that must be taken more seriously (Gezer, 2014). It is certain that in the event of a future major earthquake in Istanbul, a city with high population growth, there will be a significant loss of life and property. This is supported by the experiences of the various destructive earthquakes that have occurred in cities near Istanbul over the past century. The 1967-Adapazarı, 1999-Düzce, and 1999-İzmit earthquakes were some of these earthquakes. In the 1999 İzmit-Gölcük earthquake alone, 17.435 people lost their lives, and 100.378 buildings were destroyed or severely damaged in the broader area, including Istanbul (Hoşgören, 2000). Over the ongoing course of human development, the losses of life and property in such disasters have increased, and the main cause of this is the construction resulting from population density and unplanned urbanization (Hacısalihoğlu, 2001). According to scientists' predictions, the possibility of a major earthquake along the North Anatolian fault line, upon which Istanbul lies, is high in the future. Such an earthquake in Istanbul, which is densely populated, would cause significant losses of life and property (Üstün & Anagün, 2016). Therefore, it is important to examine this situation in detail and to ensure that buildings are earthquake-resistant in the densely populated city of Istanbul.

Population growth, which is an important SSI topic for Istanbul, can be addressed in educational environments with innovative approaches. One of these innovative approaches is the PBL approach, which is based on an individual's place of residence, as described above. The aim of this study is to examine the effects of teaching the subject of population increase in Istanbul as a place-based SSI topic on the conceptual understanding of preservice science teachers. The research question of this study is as follows: Does place-based SSI instruction have a statistically significant effect on the conceptual understanding of preservice science teachers?

7.3 Implementation Procedure

The research application process lasted 8 weeks, including pre-test and post-test applications. The first week of the application was the pre-test application; the second week entailed providing information to the preservice teachers about the PBL approach, SSI, and population growth. In the third, fourth, fifth, and sixth weeks, the preservice teachers completed their research and the application processes for the problems they had identified with the PBL approach and made presentations. A general evaluation was performed in the seventh week, and the post-test application was carried out in the eighth week.

Week 1 Conceptual understanding questions were applied to preservice science teachers as a pre-test.

Week 2 The preservice teachers were informed about the PBL approach, SSI, and population growth. They were asked the following question: "What are the effects of population growth in Istanbul?" A class discussion was held about the factors they mentioned, including earthquake probability, air pollution, water resources, and noise pollution, and four problem situations were identified. The class was divided into four groups, and each group chose one of the identified problem situations to investigate the effects on Istanbul with the PBL approach. To ensure that the groups were not too large, two different groups explored each topic. An example process was provided for each topic. Each group consisted of six people. The application process, including the number of groups, the topics each group investigated, and the research questions they formulated based on the problem situations they identified, is shown in Table 7.1.

Table 7.1 Implementation process

Week	Group No.	Topic	Research question
1	Whole class	Application of pre-tests	
2	Whole class	Place-based learning approach, SSI, and population growth	
3	1	Earthquake probability	What is the relationship between population growth and earthquake-resistant housing?
	2	Earthquake probability	What is the relationship between population growth and earthquake-resistant housing?
4	3	Air pollution	What is the relationship between population growth and air pollution?
	4	Air pollution	What is the relationship between population growth and air pollution?
5	5	Water resources	What is the relationship between population growth and water resources?
	6	Water resources	What is the relationship between population growth and water resources?
6	7	Noise pollution	What is the relationship between population growth and noise pollution?
	8	Noise pollution	What is the relationship between population growth and noise pollution?
7	Whole class	General evaluation	
8	Whole class	Application of post-tests	

The following instructions were given to the preservice teachers regarding the steps to follow during the application process:

1. Conduct research on the subject and determine where you will go for the research.
2. Plan your research with your group members regarding what you will do at the selected location. Determine the questions you will ask relevant individuals.
3. Conduct interviews with relevant individuals and local residents at the location you chose to visit. Collect evidence, access documents, and take notes about your trip.
4. After returning from the site visit, prepare a presentation with your group members. The presentation should include information, evidence, and a concluding activity or material that can propose solutions to your research problem.
5. Present the information, evidence, and concluding activity or material to your classmates regarding what you learned during the site visit. Conduct the concluding activity with the other groups in the classroom.
6. Prepare a report on the effects of population growth in Istanbul based on your own research and the research of the other groups.

7. Discuss your informed opinions on the effects of population growth with the other groups in the classroom by presenting the report you prepared (adapted from Maguth & Hilburn, 2011).

Week 3 Two groups of preservice teachers conducted research on natural disasters and earthquakes in relation to the topics they chose. They studied what natural disasters and earthquakes are, how earthquakes occur, the magnitude and intensity of earthquakes, and seismic waves. Additionally, they researched measures that can be taken before, during, and after an earthquake to minimize its effects. They decided to visit a seismology center and planned to ask questions such as: "What is an earthquake?", "Why does it occur?" "Are earthquake magnitude and intensity the same?" "How long do earthquakes last, and how much can we feel them?" "What is a fault? Is there a fault in every earthquake zone?" "What measures should be taken in densely populated areas to minimize earthquake damage?" "Can you provide information on earthquake prevention designs used in different countries?" and "What should an earthquake-resistant building look like?". They visited the seismology center and asked the employees their questions. They gained knowledge on topics such as earthquakes, earthquake magnitude, earthquake intensity, and the relationship between earthquakes and faults. Additionally, they learned about measures to reduce the impact of earthquakes, such as securing household items, using innovative technologies to allow buildings to sway horizontally instead of trembling and bending during earthquakes, and constructing solid buildings in accordance with regulations. The preservice teachers implemented an activity called "Earthquake Simulation" with their classmates using the information and evidence they obtained from their research and the visit to the seismology center to provide solutions to their research questions. The aim of this activity was to determine how the use of rail or rail-free system designs during construction and the securing of household items during an earthquake affect the outcomes of earthquakes. The materials required for the activity included cardboard, cardboard boxes, scissors, glue, play-dough, staples, a silicone gun, a sponge, paint, colored A4 paper, a honey box, a needle, thread, and wheels. The activity included the following steps:

1. Preservice teachers were divided into groups and each group carried out the activity.
2. In the first part of the activity, preservice teachers designed two rooms, one with a rail system and the other without, using cardboard boxes.
3. Preservice teachers applied vibrations first to the room with the rail system and then to the room without.
4. Preservice teachers observed and recorded changes in the positions of household items as a result of the vibrations and interpreted the differences between rooms with rail and rail-free systems. This was related to the information obtained from the seismology center visit, indicating that buildings can sway horizontally during earthquakes instead of trembling and bending (Fig. 7.1).
5. In the second part of the activity, preservice teachers designed two rooms in a cardboard box, with one containing fixed items (using pins and adhesives) and the other consisting of unfixed items.

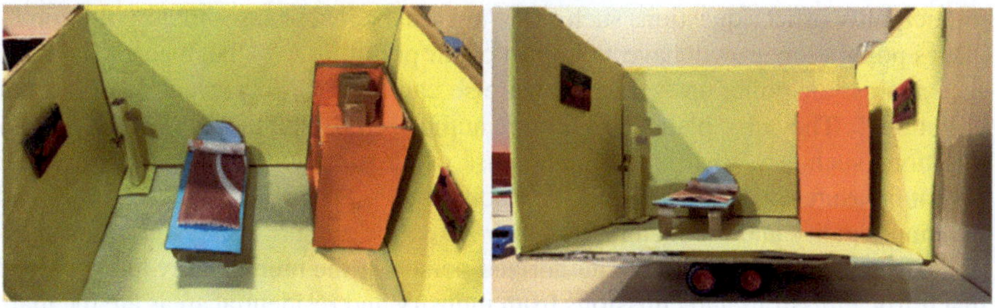

Fig. 7.1 An example of a process-oriented rail system design

Fig. 7.2 Room designs with fixed and unfixed items

6. They first applied vibrations to the fixed items and then to the unfixed items.
7. The preservice teachers observed and recorded changes in the position of the items during the vibrations and interpreted the differences between the fixed and unfixed rooms during a simulated earthquake. This was related to the issue of fixing household items based on the information gathered during the visit to the observatory (Fig. 7.2).
8. At the end of the activity, the preservice teachers concluded that fixing items and using rail systems would significantly reduce damage during an earthquake.

Week 4 Two groups of preservice teachers researched topics related to air pollution, pollutant gases, the harmful effects of air pollution, and measures that can be taken against air pollution based on the topic they chose. To better understand the relationship between population growth in Istanbul and air pollution, they examined the air quality situation in Istanbul by districts with information from the website https://havakalitesi.ibb.gov.tr/. The preservice teachers realized the importance of being informed about this issue, especially for Istanbul, the city in which they lived, and they examined the relationship between population and air pollution in the city. They decided to visit the offices of a district municipality and a vehicle parts manufacturing factory. They planned to ask the authorized person they met at the municipality offices questions such as "What are the main causes of air pollution?", "Does population growth have an impact on air pollution?" "What measures do you take against air pollution?". and "What else can be done to prevent or reduce air pollution?". They also planned to ask the authorized person they met at the factory questions such as "What is the main cause of air pollution in your factory?" and "What measures do you take against air pollution?". The preservice teachers visited the municipality and the factory, and they asked the authorized persons their questions. In their meeting with the authorized person at the municipality, they learned that traffic emissions and emissions from industry and residential areas are the main causes of air pollution. They also learned about the negative effects of residential areas with population growth on air pollution, the regulations related to air pollution caused by heating and industry, and the work the municipality does to prevent air pollution. In their meeting with the occupational health and safety specialist at the factory, they learned that the paints used in the factory affect air pollution and that measures such as ventilation, air filters, and dust collectors are applied to reduce the impact.

The preservice teachers of these groups carried out an activity titled "Which Is the Best Air Filter?" with their classmates using the information, evidence, and possible solutions they obtained from their research and visits. The aim of "Which Is the Best Air Filter?" was to determine the effects of filters made using different materials on the prevention of air pollution. The materials required for the activity included four jars, four cardboard cups, four packets of rubber bands, paper, Vaseline, cotton, filter paper, thin fabric, and a sieve. The following steps were followed in the course of the activity:

1. The preservice teachers were divided into groups, and each group conducted the activity.
2. First, the preservice teachers cut the paper into the size of the cardboard cups, glued them onto the cups, and applied Vaseline to the papers.
3. They then placed each cup into a jar and put filter paper, strainer, cotton, or thin fabric over the mouth of the jar with a rubber band.
4. They left each jar in the same external environment for an extended period of time and recorded their observations on a chart, monitoring the process outside of class.
5. The preservice teachers observed the jars for 10 days, then removed the papers from the cups and made inferences by observing which materials had more particles attached to the cups in the jars covered by each type of material (Fig. 7.3).
6. They determined that the largest amounts of particles adhering to the cups were seen in the jars covered by a strainer, cotton, thin fabric, and filter paper in that order. This was related to the use of ventilation, air filters, and dust collectors to prevent the effects of air pollution based on information obtained during research visits.
7. At the end of the activity, they concluded that the use of air filters would greatly reduce air pollution.

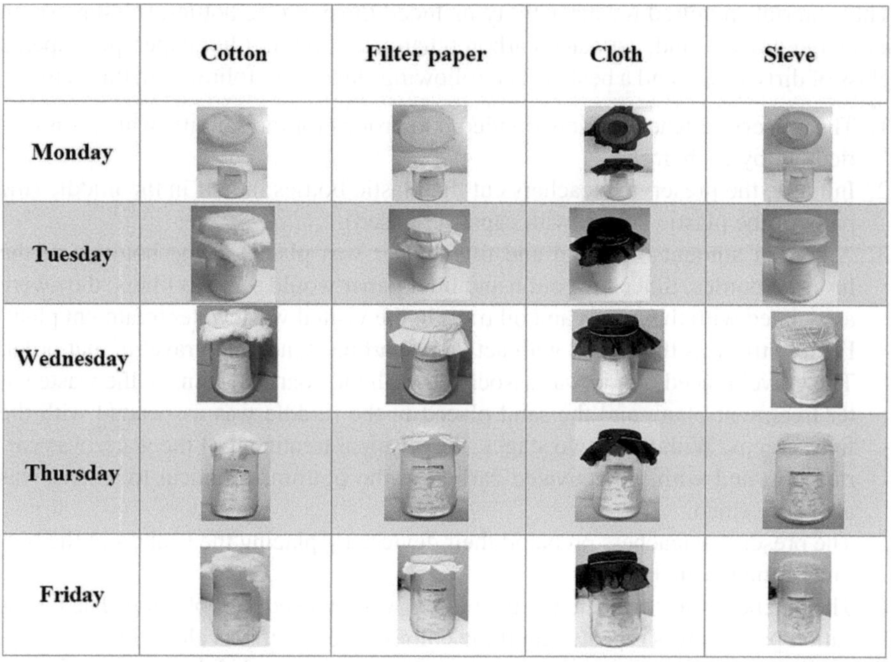

Fig. 7.3 A segment of the chart produced by the preservice teachers in the course of the activity

Week 5 This week, two groups of preservice teachers conducted research on water resources, the relationship between population growth and water resource conservation, the negative effects of water pollution, and the measures taken to prevent water pollution in Istanbul. Realizing the importance of knowledge on this subject, particularly regarding their hometown of Istanbul and the relationship between population and water resources, the preservice teachers decided to visit a wastewater treatment plant. They planned to ask the plant's authorities questions such as "Has the increasing population affected the plant?", "How is the treatment process conducted at the plant?" "What could be the future developments in water treatment if the population continues to increase?" "What can we do to protect our waters?". The preservice teachers visited the wastewater treatment plant, met with the authorities, and asked their questions. The plant treats the wastewater of approximately 3.5 million people daily, and its treatment capacity has increased due to population growth. The treatment process involves physical, chemical, and biological methods, and the preservice teachers obtained information about points to consider in order to protect our waters. Based on their research and visits to the plant, the preservice teachers conducted an activity titled "Water Filter Construction" with their classmates, which could provide a solution to their research questions. The aim of this activity was to design a water filter that would provide physical and chemical treatment to draw attention to the problem of water pollution caused by population growth.

The materials required for the activity included three plastic bottles, scissors, small and large gravel, sand, activated carbon (charcoal), cotton, filter paper, pH paper, a glass of dirty water, and a beaker. The following steps were followed in this activity:

1. The preservice teachers were divided into groups for the activity, which was carried out by each group.
2. Initially, the preservice teachers cut the plastic bottles in half in the middle (the parts of the plastic bottles with caps were used).
3. A certain amount of cotton and filter paper was placed at the bottoms of the inverted bottles. Since the cotton and filter paper would serve as filters, they were associated with the grease and oil traps in the visited wastewater treatment plant.
4. Each bottle was then filled with activated carbon, sand, and gravel in that order. The gravel placed on top was associated with the coarse screens in the wastewater treatment plant, and the sand placed in the middle was associated with the fine screens. With these two stages, the physical treatment of the water was carried out, and with the activated carbon at the bottom, chemical treatment was also performed.
5. The preservice teachers prepared their devices by placing the beakers at the bottom of the plastic bottles.
6. The purified water collected in the beaker was checked for pollution. The purification process was repeated until the color of the water was clear (Fig. 7.4).
7. At the end of the activity, the preservice teachers concluded that the use of water filters would significantly reduce water pollution and, as a result, help preserve water resources.

Fig. 7.4 An example of the design applied by the preservice teachers

Week 6 This week, two groups of preservice teachers conducted research on topics related to noise pollution, such as the definition of noise pollution, the level of noise pollution, and the effect of population density on noise pollution in some districts of Istanbul. They realized the importance of being informed on this topic, especially since they lived in Istanbul, and they examined the relationship between population growth and noise pollution. They decided to visit a sound insulation center in Istanbul to ask questions such as "What is sound insulation?", "What are sound insulation materials?", "How do sound insulation materials provide insulation?" and "Why should sound insulation be used?". The preservice teachers visited the sound insulation center, met with authority, and directed their questions to that individual. They learned about the definition of sound insulation, what sound insulation materials are, the importance of sponge-based materials in providing insulation, and the necessity of sound insulation to remove unwanted sounds from the environment. The preservice teachers then implemented an activity titled "Sound Insulation" with their classmates, which could offer a solution to their research questions based on the information and evidence they obtained from their research and visit to the sound insulation center. The aim of "Sound Insulation" was to determine how the use of sound insulation materials affects sound intensity. The materials required for the activity included insulation material (egg-patterned sponge), two cardboard boxes of equal size, a sound device (smartphone), and an app for decibel measurements. The following steps were applied in the activity:

1. The preservice teachers were divided into groups, and each group carried out the activity.
2. They attached insulation material (Fig. 7.5) to one of the two cardboard boxes of equal size and applied no treatment to the other.
3. A sound-emitting device was placed inside the cardboard boxes in sequence, and the preservice teachers were asked to compare the sound coming from the insulated box with the sound coming from the uninsulated box.
4. An app for decibel measurement was used to make the comparison.
5. The values obtained from the app were recorded, and it was determined that the sound of lower decibels came from the cardboard box with an egg-patterned sponge for insulation.

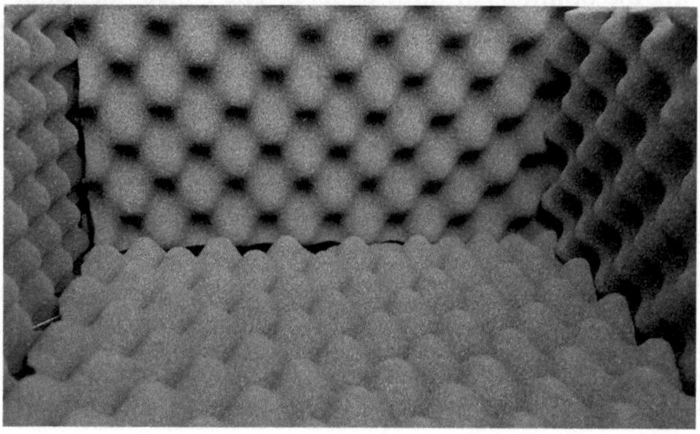

Fig. 7.5 View of the insulated box

6. At the end of the activity, the preservice teachers concluded that the use of sound insulation would greatly reduce noise pollution and, relatedly, that the reduction of noise levels in densely populated areas could be largely achieved through insulation materials, which was associated with the information obtained from the visit to the sound insulation center, indicating the importance of sponge-based materials in sound insulation.

Week 7 This week, the preservice teachers presented their reports in the classroom regarding the effects of population growth on Istanbul in the context of earthquakes, water resources, air pollution, and noise pollution using the information they obtained from their own research and the research of the other groups. They also discussed their views on the effects of population growth as conscious citizens. As a result of these discussions, the preservice teachers made inferences about the effects of population growth in Istanbul through place-based SSI instruction.

Week 8 This week, conceptual understanding questions were administered to the preservice teachers as a post-test.

7.3.1 Research Design

This research was conducted using a pre-test/post-test experimental design without a control group. A pre-test is administered before the experimental procedures in such designs without control groups. After the experimental procedures are concluded, the same test is administered again as a post-test. The effect of the experimental treatment is then determined by comparing the pre-test and post-test scores (Sönmez & Alacapınar, 2013).

The research group comprised 43 preservice science teachers enrolled in the third year of study in the faculty of education of a public university in Istanbul. While 34 of the participants were female, nine were male.

Conceptual understanding questions were used as the data collection tools in this research. The researchers developed the questions and finalized them after obtaining the opinions of three experts. The conceptual understanding questions included 15 open-ended questions intended to identify the changes in the preservice teachers' conceptual understanding. Some examples of these conceptual understanding questions are as follows: "What is population growth?" "What are the effects of population growth?" and "Is population growth a positive or negative situation for a region? Explain this with reasons."

Responses to the conceptual understanding questions administered as a pre-test and post-test were evaluated as correct, partially correct, or incorrect. Correct answers were given 2 points, partially correct answers were given 1 point, and incorrect answers were given 0 points. The quantitative scores obtained from the tests were analyzed using the IBM SPSS Program.

7.3.2 Outcomes

The findings of this study, which investigated the effect of place-based SSI instruction on the conceptual understanding of preservice teachers, were first examined to see whether the data showed normal distribution or not. In cases where the sample size is 30 or more, the normality of the data can be tested with the Kolmogorov-Smirnov test. If the significance of that test is at the 5% level ($p < 0.05$), it means that the data are not normally distributed (Gürbüz & Şahin, 2018). Since the study group of our research was larger than 30, the normality of the responses to the conceptual understanding questions was examined using the Kolmogorov-Smirnov test, and the results obtained are presented in Table 7.2.

As seen in Table 7.2, the distribution of the pre-test scores was normal ($p > 0.05$), while the distribution of the post-test scores was not normal ($p < 0.05$). Therefore, a non-parametric test, the Wilcoxon signed-rank test, was applied for data analysis. The results of the Wilcoxon signed-rank test for the scores obtained by the preservice teachers are presented in Table 7.3.

The Wilcoxon signed-rank test, a non-parametric test, was applied for the analysis of the data since the post-test scores did not show a normal distribution ($p < 0.05$) as determined by the Kolmogorov-Smirnov test results provided in Table 7.2. The results of the Wilcoxon signed-rank test for the scores of the preservice teachers are presented in Table 7.3, indicating a significant difference between the preservice teachers' pre-test and post-test scores in terms of their conceptual understanding of the topic ($Z(43) = -5.718$, $p < 0.05$). Considering the rank means and totals of the differences in scores, it is observed that the difference was in favor of positive ranks, which indicates that the preservice teachers had higher scores on the post-test compared to the pre-test. This finding demonstrates that place-based SSI instruction positively influenced the preservice science teachers' conceptual understanding.

Table 7.2 Results of the Kolmogorov-Smirnov test

Test	Statistic	df	p
Pre-test	0.97	43	0.200
Post-test	0.176	43	0.002

Table 7.3 Wilcoxon test results regarding comparisons of pre-test and post-test scores

	N	Rank average	Row sum	Z	p
Negative order	0[a]	0.00	0.00	−5.718[a]	0.000
Positive order	43[b]	22.00	946.00		
Equal	0[c]				
Total	43				

[a]Post-test < pre-test, [b]post-test > pre-test, [c]post-test = pre-test

In addition to their responses to the conceptual understanding questions in the post-test, the preservice teachers were also asked to reflect on the process. They stated that the applied approach facilitated their learning, and they had positive opinions about this instruction. They said they felt good and enjoyed the process, particularly highlighting that the field trips contributed to the deepening of their learning. Furthermore, it was also observed that the preservice teachers' conceptual understanding improved throughout the process. They worked actively and collaboratively during the implementation, and they were motivated to participate in the process.

7.4 Transfer to the Global Context

The first significant increase in the world's population occurred about 12.000 years ago when humans transitioned to an agricultural lifestyle, and the world's population was 5 million people with a population growth rate of 0.002%. Compared to the beginning of humanity, the population growth rate in the 1750s was approximately 150 times higher. Since then, it is seen that the world population has increased rapidly in parallel with the rapid increase in the annual growth of the population (Gündüz et al., 2005). Currently, the average global population growth rate is 1.7%, and if it continues to grow at the same rate, the world's population will double in approximately 41 years. The world's population is estimated to reach 30 billion by 2075 (Çamurcu, 2005).

While rapid population growth leads to various problems, such as overpopulation, a very low or decreasing population also threatens the existence and future of countries. For this reason, different population policies are implemented in various parts of the world. In this context, setting and implementing goals that can respond to different needs is necessary. However, this is not always easy for every country. Currently, there are three main types of population policies generally implemented:

1. Population policies aimed at reducing the population growth rate, such as those implemented in countries like China and India.
2. Population policies aimed at increasing the population growth rate by countries with rapidly declining populations, such as Sweden, Finland, and Denmark.
3. Population policies aimed at improving the quality and quantity of the population as implemented by developing countries such as Türkiye (Doğan, 2011).

In today's educational processes, it is seen that the effects of different sociocultural environments vary. It is inevitable that countries with different population densities also have different sociocultural structures. Considering the population policies implemented to reduce population growth rates, increase population growth rates, and improve the quality and quantity of the population, educational environments should also be shaped according to the needs of individual countries. If a country's population is increasing or decreasing, the students living in that country should be expected to discuss the consequences of the situation in their educational environments and, if necessary, engage in problem-solving efforts.

The importance of teacher education has been emphasized from the beginning of research in this field. A preservice teacher who receives a high-quality education can help students structure their knowledge by using more diverse methods and techniques at the beginning of his or her career. They can also help students make useful inferences about social issues. In this study, the steadily growing population of the city of Istanbul was taken as an example of an SSI. The participating preservice teachers themselves determined the "places" where research related to Istanbul's population growth would be conducted with institutions for a PBL approach. In other areas where similar situations are experienced, preservice teachers could conduct research to solve the problems they identify and pursue place-based SSI learning by attending various institutions and organizations. The use of place-based learning approaches can be recommended in countries with different rates of population growth or decline.

Although the paths followed in teacher education may vary, the issue of population growth is a topic that concerns every country. Therefore, opportunities should be provided for learners to address this issue as responsible citizens. Educational environments can be diversified by allowing learners to use different skills and actively participate in learning. In this research, different places were visited, and laboratory environments were also used. Preservice teachers carried out various activities in cooperation in those environments. They supported their learning processes by conducting research, designing experimental processes, presenting their work, implementing activities together, and completing peer/self-assessments. There is also a need for such differentiation and diversification in other sociocultural environments. STEM, inquiry-based, problem-based, and project-based learning approaches can be integrated with PBL processes. In addition, place-based learning approaches can be included in teacher education programs with modern applications such as virtual and augmented reality applications, robotic coding activities, and design skill workshops, which are emphasized as necessities of education systems highlighting 21st-century skills.

7.5 Conclusion

In this research, preservice teachers recognized that knowledge gained from an external institution can be transferred to the classroom environment following contact with that institution. This study has determined that place-based SSI instruction positively affected the conceptual understanding of preservice teachers by enabling them to acquire knowledge in Istanbul within the context of population growth. Additionally, it was observed that the opinions of preservice teachers regarding the implementation were positive.

References

Adams, A. E., Miller, B. G., Saul, M., & Pegg, J. (2014). Supporting elementary pre-service teachers to teach STEM through place-based teaching and learning experiences. *The Electronic Journal for Research in Science & Mathematics Education, 18*(5). http://ejse.southwestern.edu/article/view/12958

Akgün, A., Gönen, S., & Yılmaz, A. (2005). Fen bilgisi öğretmen adaylarının karışımların yapısı ve iletkenliği konusundaki kavram yanılgıları [Misconceptions of preservice science teachers regarding the structure and conductivity of mixtures]. *Hacettepe Üniversitesi Eğitim Fakültesi Dergisi, 28*(28), 1–8. http://efdergi.hacettepe.edu.tr/shw_artcl-751.html

Bachtold, M. (2013). What do students 'construct' according to constructivism in science education? *Research in Science Education, 43*(6), 2477–2496. https://doi.org/10.1007/s11165-013-9369-7

Bayat, D. (2010). *Gelişmekte olan ülkelerde nüfus sorunu ve Türkiye örneği* [Population problem in developing countries and the example of Türkiye]. Unpublished doctoral dissertation. Marmara Üniversitesi.

Bozkurt, B. (2013). *Kürtaj yasalarının nüfus politikalarıyla ilişkisi: 1960'lardan günümüze Türkiye* [Relationship between abortion laws and population policies: Türkiye from the 1960s to the present]. Unpublished undergraduate graduation thesis. Yıldız Teknik Üniversitesi.

Çamurcu, H. (2005). Dünya nüfus artışı ve getirdiği sorunlar [The increase of the world population and the problems it causes]. *Balıkesir Üniversitesi Sosyal Bilimler Enstitüsü Dergisi, 8*(13), 87–105. https://dergipark.org.tr/en/pub/baunsobed/issue/50339/651864

Cooperation Project for Environment, Climate and Health. (2022). *Hava kirliliğinin sağlık etkileri* [Health effects of air pollution]. https://www.env-health.org/wp-content/uploads/2022/03/Hava_Kirliligi_Bilgi_Notu.pdf. Reached on 17 Apr 2023.

Cunningham, R. T., & Turgut, M. F. (1996). *İlköğretim fen bilgisi öğretimi* [Elementary science teaching]. Yök/Dünya Bankası.

Darmofal, D. L., Soderholm, D. H., & Brodeur, D. R. (2002). Using concept maps and concept questions to enhance conceptual understanding. *Proceedings - Frontiers in Education Conference*. https://doi.org/10.1109/FIE.2002.1157954

Dawson, V. (2015). Western Australian high school students' understandings about the socioscientific issue of climate change. *International Journal of Science Education, 37*(7), 1024–1043. https://doi.org/10.1080/09500693.2015.1015181

Demirel, R. (2016). Argümantasyon destekli öğretimin öğrencilerin kavramsal anlama ve tartışma istekliliklerine etkisi [The effect of scientific argumentation students' conceptual understanding and willingness for argumentation]. *Kastamonu Eğitim Dergisi, 24*(3), 1087–1108. https://dergipark.org.tr/en/download/article-file/210077

Doğan, M. (2011). Türkiye'de uygulanan nüfus politikalarına genel bakış [General overview of the population policy in Turkey]. *Marmara Coğrafya Dergisi, 23*, 293–307. https://dergipark.org.tr/en/download/article-file/3244

Doğan, S. (2015). *Cumhuriyet döneminde Türkiye nüfusunda meydana gelen değişimlerin okul öncesi eğitim sistemine yansıması* [The reflection of the changes in the Turkish population in the Republican period to the pre-school education system]. Unpublished doctoral dissertation. Uludağ Üniversitesi.

Dubel, M., & Sobel, D. (2008). Place-based teacher education. In D. A. Gruenewald & G. A. Smith (Eds.), *Place-based education in the global age: Local diversity* (pp. 309–344). Lawrence Erlbaum Associates.

Ebersohn, L., & Ferreira, R. (2012). Rurality and resilience in education: Place-based partnerships and agency to moderate time and space constraints. *Perspectives in Education, 30*(1), 30–42. https://journals.ufs.ac.za/index.php/pie/article/view/1730/1704

Es, M., & Ateş, H. (2010). Kent yönetimi, kentlileşme ve göç: Sorunlar ve çözüm önerileri [Urban management, urbanization and migration: Problems and solutions]. *Journal of Social Policy Conferences, 10*(48), 205–248. https://dergipark.org.tr/en/pub/iusskd/issue/890/9892

Falkenmark, M., Lundqvist, J., & Widstrand, C. (1989). Macro-scale water scarcity requires micro-scale approaches: Aspects of vulnerability in semi-arid development. *Natural Resources Forum, 13*(4), 258–267. https://doi.org/10.1111/j.1477-8947.1989.tb00348.x

Gezer, İ. (2014). Kentsel dönüşüm; Ne, niçin, nasıl? [Urban transformation; what, why, how?]. *Kamuda Sosyal Politika*, 41–48.

Gobert, J. D., & Clement, J. J. (1999). Effects of student-generated diagrams versus student generated summaries on conceptual understanding of causal and dynamic knowledge in plate tectonics. *Journal of Research in Science Teaching, 36*(1), 39–54. https://doi.org/10.1002/(SICI)1098-2736(199901)36:1<39:AID-TEA4>3.0.CO;2-I

Gödek, Y., & Polat, D. (2017). Fen eğitiminde kavram öğretimi [Concept teaching in science education]. In M. P. Demirci Güler (Ed.), *Fen bilimleri öğretimi* [Science teaching] (p. 48). Pegem Akademi Yayıncılık.

Görmez, K. (2015). *Çevre sorunları* [Environmental problems] (3rd ed.). Nobel Yayınları.

Green, A. (2012). *The integration of engineering design projects into the secondary science classroom*. Unpublished master's thesis. Michigan State University.

Gündüz, M., Kutlar, A., Gündüz, F., & Esgin, A. (2005). *Nüfus sorunu. Eğitim ve sosyo-ekonomik dönüşüm bağlamında Türkiye'de doğurganlık* [Population problem. Fertility in Turkey in the context of education and socio-economic transformation]. Anı yayıncılık.

Gürbüz, S., & Şahin, F. (2018). *Sosyal bilimlerde araştırma yöntemleri: Felsefe, yöntem, analiz* [Research methods in social sciences: Philosophy, method, analysis]. Seçkin Yayıncılık.

Hacısalihoğlu, İ. Y. (2001). *Türkiye'nin kentsel gelişme süreci ve 1999 Marmara depremi mekânsal planlamadan kopmanın acı bilânçosu* [Turkey's urban development process and the 1999 Marmara earthquake, the bitter balance of the break from spatial planning]. Çantay Yayıncılık.

Hoşgören, M. Y. (2000). İstanbul ve deprem [Istanbul and earthquake]. *Türk Coğrafya Dergisi, 35*, 1–24. https://dergipark.org.tr/en/download/article-file/198654

İnel Ekici, D. (2014). Kavram öğretimi [Concept teaching]. In Ş. S. Anagün, & N. Duban (Eds.), *Fen bilimleri öğretimi* [Science teaching] (1st ed.) Anı Yayıncılık.

Istanbul Metropolitan Municipality. (2001). *Sayım yıllarına göre İstanbul nüfusu, sayımlar arası yıllık nüfus artış hızı ve yıl ortası nüfus tahmini* [Istanbul population by census years, annual population growth rate between censuses and mid-year population estimation]. http://www.ibb.gov.tr/trTR/BilgiHizmetleri/Istatistikler/Documents/demografi/t211.pdf. Reached on 17 Apr 2023.

Jennings, N., Swidler, S., & Koliba, C. (2005). Place-based education in the standards-based reform era-conflict or complement? *American Journal of Education, 112*(1), 44–65. https://doi.org/10.1086/444522

Kong, S. C., Cheung, W. M. Y., & Zhang, G. (2023). Evaluating an artificial intelligence literacy programme for developing university students' conceptual understanding, literacy, empowerment and ethical awareness. *Educational Technology & Society, 26*(1), 16–30. https://doi.org/10.30191/ETS.202301_26(1).0002

Köse, S., Coştu, A. G. B., & Keser, A. G. Ö. F. (2003). Fen konularındaki kavram yanılgılarının belirlenmesi: Tga yöntemi ve örnek etkinlikler [Determination of students' misconceptions in science: Activities through POE method]. *Pamukkale Üniversitesi Eğitim Fakültesi Dergisi, 13*(13), 43–53. https://dergipark.org.tr/en/download/article-file/114819

Kutsal, S., & Polatoğlu, Ç. (2023). Türkiye'de artan konut ihtiyacı ve konut sorununun dinamikleri [Increasing housing need and dynamics of housing problem in Turkey]. *Topkapı Sosyal Bilimler Dergisi, 2*(1), 133–158. https://dergipark.org.tr/en/pub/tjssd/issue/75780/1178913

Maguth, B. M., & Hilburn, J. (2011). The community as a learning laboratory: Using place-based education to foster a global perspective in social studies. *Ohio Social Studies Review, 47*(1), 27–34.

Menteşe, S. (2017). Çevresel sürdürülebilirlik açısından toprak, su ve hava kirliliği: Teorik bir inceleme [Soil, water and air pollution in terms of environmental sustainability: A theoretical review]. *Uluslararası Sosyal Araştırmalar Dergisi, 10*(53), 381–389. https://doi.org/10.17719/jisr.20175334127

Morillas, J. M., Escobar, V. G., Sierra, J. A. M., Gomez, R. V., & Carmona, J. T. (2002). An environmental study in the city of caceras (Spain). *Applied Acoustics, 63*, 1061–1070. https://doi.org/10.1016/S0003-682X(02)00030-0

Morris, C. G. (1996). *Understanding psychology*. Prentice Hall.

Novak, J. D. (2010). *Learning, creating, and using knowledge: Concept maps as facilitative tools in schools and corporations*. Routledge.

Özyonar, F., & Peker, Y. (2008). Sivas kent merkezindeki çevresel gürültü kirliliğinin araştırılması [Investigation of environmental noise pollution in Sivas city center]. *Ekoloji, 18*(69), 75–80.

Senemoğlu, N. (2001). *Kuramdan uygulamaya gelişim ve öğrenme* [Development and learning from theory to practice]. Gazi Kitabevi.

Sinan, O. (2007). *Fen bilgisi öğretmen adaylarının proteinler ve protein sentezi ile ilgili kavramsal anlamaları* [Conceptual understanding of pre-service science teachers about proteins and protein synthesis]. Unpublished doctoral dissertation. Balıkesir Üniversitesi.

Sofuoğlu, A. (2003). *Hava kirliliği* [Air pollution]. İzmir Yüksek Teknoloji Enstitüsü.

Sönmez, V., & Alacapınar, F. (2013). *Örneklendirilmiş bilimsel araştırma yöntemleri* [Illustrated scientific research methods] (1st ed.). Anı Yayıncılık.

Stickney, J. (2020). Philosophical walks as a place-based environmental education. *Journal of Philosophy of Education, 54*(4), 1071–1086. https://doi.org/10.1111/1467-9752.12469

Topal, K. (2011). Çevre sorunlarının kaynağı olarak nüfus artışı mı? Tüketim mi? Neo-Malthusyen düşünceye eleştirel bir yaklaşım [Population growth as the source of environmental problems? Is it consumption? A critical approach to neo-Malthusian thought]. *Türk İdare Dergisi, 470*, 133–152.

Turkish Statistical Institute. (2022). *Nüfus istatistik verileri* [Population statistics data]. http://www.tuik.gov.tr. Reached on 15 Mar 2023.

Türkyılmaz, A., (2010). *Dünyada ve ülkemizde su* (Su yönetimi ve mevzuatı) [Water in the world and in our country (Water management and legislation)]. Sarıyıldız Ofset.

United Nations. (2022). *World population prospects 2022: Summary of results* (UN DESA/POP/2021/TR/NO.3). United Nations Department of Economic and Social Affairs Population Division. https://www.un.org/development/desa/pd/sites/www.un.org.development.desa.pd/files/wpp2022_summary_of_results.pdf. Reached on 15 Mar 2023.

Urban Transformation Research Report. (2014). *Türkiye, Cushman & Wakefield araştırma yayını*. http://www.cushmanwakefield.com. Reached on 17 Apr 2023.

Üstün, A. K., & Anagün, A. S. (2016). İstanbul un ilçelerinin afet yönetimi açısından önem ağırlıklarının Analitik Hiyerarşi Süreci kullanılarak belirlenmesi [Determining the importance weights of the districts of Istanbul in terms of disaster management using the Analytical Hierarchy Process]. *Gazi Üniversitesi Mühendislik-Mimarlık Fakültesi Dergisi, 31*(1), 119–128. https://doi.org/10.17341/gummfd.08342

Watts, G. R. (1995). A comparison of noise measures for assessing vehicle noisiness. *Journal of Sound and Vibration, 180*(3), 493–512. https://doi.org/10.1006/jsvi.1995.0092

World Health Organization. (2022). *Air quality database*. https://www.who.int/data/gho/data/themes/air-pollution/who-air-quality-database. Reached on 17 Apr 2023.

Yüksek, T. (2004). Türkiye'nin su kaynakları ve havza planlamasına dönük genel değerlendirmeler [General evaluations on Turkey's water resources and basin planning]. *Kafkas Üniversitesi Artvin Orman Fakültesi Dergisi, 1*(2), 71–83.

Zannin, P. H. T., Diniz, F. B., & Barbosa, W. A. (2002). Environmental noise pollution in the city of Curitiba (Brazil). *Applied Acoustics, 63*(4), 351–358. https://doi.org/10.1016/S0003-682X(01)00052-4

Chapter 8
Enhancing Preservice Teachers' Understanding and Application of Place-Based Socioscientific Issues: A Focus on Renewable Energy and Perspective-Taking

Nilay Muslu and Hakan Isik

8.1 Introduction

Students in the twenty-first century are expected to be educated and responsible citizens who can make informed decisions. To support this aim, there is a shift in science education towards emphasizing contextualized scientific knowledge to address complex and ill-structured SSI rather than memorizing scientific facts (Zeidler et al., 2005). SSI are complex and controversial scientific issues that affect societies at personal, regional, or global levels and address the cultural, political, or moral aspects of these issues (Lee et al., 2013; Sadler, 2004; Sadler & Zeidler, 2005; Zeidler, 2014). Incorporating SSI into teaching is essential for student learning. It enables students to improve their science knowledge, increase their awareness both socially and environmentally, assist them in evaluating the problems critically, and help them to become informed citizens who are sensitive to the problems in their environment (Herman, 2018; Pouliot, 2008; Zeidler et al., 2005). As Herman et al. (2022) discussed, several researchers have outlined the essential components of effective instruction in SSI. These components include constructing instruction around a captivating, pertinent, and authentic issue; giving early emphasis to the issue in the instruction; providing students with a culminating experience tied to the issue; and encouraging higher-order practices such as scientific argumentation, reasoning, and decision-making through scaffolded experiences and deliberate pedagogical support, which helps students engage with the issue from multiple perspectives (Herman, 2018; Presley et al., 2013; Sadler, 2009).

N. Muslu (✉) · H. Isik
Department of Science Education, Faculty of Education,
Mugla Sitki Kocman University, Muğla, Türkiye
e-mail: nilaymuslu@mu.edu.tr; hisik@mu.edu.tr

B. Namdar, E. Karahan (eds.), *Socioscientific Issues Focused Teacher Education*, Contemporary Trends and Issues in Science Education 60,
https://doi.org/10.1007/978-3-031-55233-5_8

8.1.1 Perspective Taking

Sadler et al. (2007) argued that SSI learning experiences should aid students in not only understanding science but also in coping with complex issues. Thus, they proposed socioscientific reasoning (SSR) as a way to describe the reasoning practices involved in negotiating and solving SSI. Perspective-taking is one of the dimensions of SSR (Sadler et al., 2007). Perspective-taking is defined as a person's ability to recognize and consider other points of view while addressing a problem. Students must be aware of those views and open to engaging in productive discourse and decision-making on complex issues. For this reason, this skill should be considered during SSI participation when the teacher facilitates argumentation for decision-making and develops functional scientific literacy (Leung & Cheng, 2020; Kahn & Zeidler, 2016; Sadler et al., 2007). In this chapter, we aim to look at the importance of integrating perspective-taking in SSI.

Leung et al. (2015) argue that to evaluate science effectively, we must consider social and other dimensions as well as scientific perspectives. In other words, multiple perspectives are necessary for a holistic evaluation of science. Choi and Lee (2021) argued that it is important for students to understand the complexity of human elements in science. They suggested that students explore different aspects of science, including its role in society and how it is communicated both inside and outside of classrooms. Incorporating perspective-taking into teaching will help students develop social and environmental awareness, assess the challenges they face with a critical mind, and become educated citizens who wonder about the problems around them (Pouliot, 2008; Zeidler et al., 2005). Improving perspective-taking skills will help students improve their skills in leadership, ethics, problem-solving, complicated decision-making, critical thinking, and reflective practices (Kahn & Zeidler, 2016; Kinskey & Zeidler, 2021; Newton & Zeidler, 2020). In short, perspective-taking is a crucial skill for students to master to become educated citizens who can actively participate in and resolve SSI (Newton & Zeidler, 2020; Kahn & Zeidler, 2019; Sadler et al., 2007; Zeidler et al., 2005).

8.1.2 Perspective-Taking for Place-Based SSI

The literature has emphasized the importance of perspective-taking for place-based SSI as an essential skill for meaningful participation and eventual SSI resolution (Kahn & Zeidler, 2019; Newton & Zeidler, 2020; Sadler et al., 2007; Zeidler et al., 2005). Utilizing perspective-taking through direct interaction or observation enables students to develop a comprehensive understanding of the environment and easily grasp how these issues may affect others. According to research on place-based SSI in environmental education, perspective-taking skills are crucial for developing empathy and applying emotive reasoning (Newton & Zeidler, 2020; Herman et al., 2021; Herman et al., 2020). The cultivation of perspective-taking and emotive reasoning plays a vital role in moral development and is integral to the progress of SSR, scientific literacy, and environmental responsibility and management (Herman, 2018; Kahn & Zeidler, 2016; Sadler & Zeidler, 2005). In a place-based course, learners immerse themselves in a problem in their environment with a holistic view through field experiences and interactions with diverse groups of people who are affected by the problem. Therefore, they will learn to focus on the natural sciences, the human sciences, and different perspectives (Herman et al., 2020).

Resolving SSI effectively requires individuals to internalize and empathize with the perspectives and emotions of those grappling with the moral dilemmas connected to these issues. This process can shift students' perceptions from a belief in their superiority over nature and reliance solely on technology to a more responsible stance. Moreover, previous research indicates that individuals' perspectives on the nature and limitations of science are often linked to their personal responsibility and commitment to taking action in addressing SSI (Herman, 2015, 2018; Herman et al., 2020). By fostering such perspective-taking and empathy, students can develop strong connections to the natural world (Herman et al., 2018; Kurdyavtsev et al., 2012) and enhance their willingness to engage in proactive measures regarding the issue (Choi & Lee, 2021; Mayer et al., 2009). The implementation of place-based SSI educational approaches is crucial for nurturing the desired relationship between nature and culture among students. Traditional science classrooms often fail to connect students with the real-world challenges experienced by their communities (Bang & Marin, 2015). Furthermore, place-based SSI allows students to develop a sense of attachment, leading to more student engagement (Kim et al., 2020).

In conclusion, perspective-taking is crucial for analyzing complex SSI, fostering functional scientific literacy, developing empathy, and facilitating moral decision-making. Incorporating place-based SSI learning experiences can help students develop strong affective connections to nature as well as a deeper understanding of the social, cultural, and environmental factors that influence SSI. Engaging in perspective-taking helps improve higher-order thinking skills, metacognitive thinking, and reflective practices. This understanding will allow them to engage in meaningful decision-making and position-taking.

8.1.3 Preservice Teacher Education in the Context of SSI-Based Teaching and Learning

In Türkiye, the teaching of SSI took its place in the curriculum of the Ministry of National Education as a dimension of science, technology, society, and environment learning areas in 2013 (Ministry of National Education [MoNE], 2013). However, studies have shown that traditional methods are widely used in science lessons (Apaydın & Kandemir, 2018; Okur Akçay et al., 2016), as teachers do not have sufficient knowledge and experience in teaching SSI (Aydın et al., 2021; Han-Tosunoğlu

& İrez, 2017). Also, teachers believe teaching SSI is a challenging task due to its complex and ill-structured structure (Saunders & Rennie, 2013). Previous research found a variety of reasons for the limited implementation of SSI in science courses, such as teachers' avoidance of the sense of control (Lee & Yang, 2019), teachers' lack of knowledge on the nature of science, sociological, economic, political, and ethical dimensions (Evren Yapıcıoğlu & Kaptan, 2018). In addition, there is a lack of classroom management skills during student-centered SSI-based teaching (Hancock et al., 2019) and limited SSI-based materials and assessment practices (Hancock et al., 2019; Lee et al., 2006). Time constraints also pose a challenge, with limited time for planning SSI and limited facilitated instruction (Evren Yapıcıoğlu & Kaptan, 2018). Moreover, there is a lack of support from the parents, as they often do not view SSI as a part of the actual teaching (Evren Yapıcıoğlu & Kaptan, 2018), and there is limited support from the administrators (Saunders & Rennie 2013). In addition to these, since students are accustomed to viewing their instructors as a source of knowledge, they also face difficulties adjusting to their new roles in this new culture (Lee & Yang, 2019).

Preservice teachers also have concerns related to integrating SSI-based instruction into their future classrooms. Preservice science teachers do not have sufficient self-efficacy beliefs to integrate SSI into their teaching (Lee et al., 2006; Sibic & Topçu, 2020). These beliefs are affected by their limited teaching practices in real classroom settings, limited content and pedagogical knowledge, and lack of experience in teaching SSI (Lee et al., 2006; Sibic & Topçu, 2020). Similar to in-service teachers, preservice teachers also have difficulties with classroom management and time management for planning and preparing SSI-based lessons (Sibic & Topçu, 2020). Preservice teachers are also concerned about students' academic abilities to discuss controversial SSI topics (Borgerding & Dagistan, 2018). In order to minimize these negativities, preservice teachers should have sufficient knowledge about SSI and activity planning, organizing discussion activities that can be used in this activity, and evaluating the activities, and they should benefit from expert help and support in this planning and implementation process (Reis & Galvao, 2009). For this reason, when preparing preservice science teachers to use SSI curricula, it is crucial for educators to focus on essential skills in planning and implementing the SSI (Sibic & Topçu, 2020). To be able to achieve this aim, opportunities for preservice teachers to plan and integrate SSI in different teaching settings, both real and microteaching, could help them improve their self-efficacy beliefs in integrating SSI into their lessons (Sibic & Topçu, 2020). Another way to support preservice teachers is through field experiences where they can gain contextual experiences in the classroom to allow opportunities to gain a deeper pedagogical understanding of SSI and improve their skills associated with pedagogy and practice (Kinskey & Zeidler, 2021). Moreover, incorporating a community of practice where preservice teachers can exchange ideas and support each other also helps them in SSI-based instruction (Kinskey, 2020; Kinskey & Zeidler, 2021).

In line with these concerns, research has focused on how preservice teachers' views and skills evolve after completing SSI-based programs or courses, particularly their understanding of the nature of science (NOS), argumentation skills, and global citizen values. Research indicates a positive improvement in their NOS understanding, content knowledge, and argumentation skills following SSI-based instructions (Brocos & Jiménez-Aleixandre, 2020; Evren-Yapıcıoğlu & Kaptan, 2018; Herman et al., 2022; Kutluca & Aydin, 2017). However, challenges persist in implementing SSI in instruction. While changing NOS views among elementary preservice teachers has shown success, guiding them to implement NOS instruction remains a challenge (Kutluca & Aydin, 2017). Similarly, Aydın et al. (2021) stated that although teachers recognize the effectiveness of high-level thinking skills, such as argumentation, in students' learning of SSI, they do not fully realize that these activities can improve students' science knowledge. Consequently, preservice teachers struggle to understand the purpose of teaching SSI.

Another line of research explores the character and values of preservice teachers as an important aspect to consider in their integration of SSI. Lee et al. (2012) examined the character and values of preservice science teachers in South Korea as global citizens, focusing on ecological worldview, socioscientific accountability, and social and moral compassion. The findings revealed that while the teachers exhibited key elements of character and values, they had difficulties in consistently applying moral principles and had limited global perspectives. The study suggests that SSI programs can enhance SSR by promoting the recognition of SSI complexity, considering multiple perspectives, and fostering skepticism towards information.

In order to support preservice teachers in integrating SSI effectively into their future classrooms, educators must be aware of the distinctive perspectives and beliefs associated with various subject disciplines. Furthermore, Barrett and Nieswandt (2010) emphasized the importance of subject-specific identities in influencing preservice teachers' beliefs about SSI inclusion in their teaching practices. This demonstrates how critical it is to take these subject-specific identities into account during preservice education, as it can help shape their attitudes toward SSI integration.

In a related study, Newton and Zeidler (2020) found that preservice teachers' perspective-taking abilities shifted through an environmental science course designed for future K-8 teachers. Reading, researching, interacting with guest speakers, and participating in group presentations enable preservice teachers to gain a holistic view of the issue, increase their factual knowledge, and develop an awareness of its complexity. The findings showed that participating in an SSI experience with a perspective reading and writing intervention improved preservice teachers' ability to spot changes in their perspective-taking skills. The findings support the idea that explicit implementation of perspective-taking interventions can increase awareness of perspectives and enhance cognitive flexibility and reflection (Kahn & Zeidler, 2016).

Additionally, the incorporation of place-based education and SSI allows preservice teachers to analyze students' backgrounds and local settings, enabling them to design more relevant lesson plans (Nargund-Joshi, 2020). This approach focuses on students' strengths and enhances the accessibility and relevance of science content (Semken & Freeman, 2008). Brocos and Jiménez-Aleixandre (2020) highlighted the importance of involving preservice teachers in relevant and personally meaningful contextual problems to educate them as environmentally conscious individuals while also respecting and preserving their cultural traditions and local perspectives. Language barriers and limited knowledge of students' cultural and social aspects further compound these challenges. To address this, continuous dialogue about contextualization and strategies to gather more background knowledge about students' home cultures and practices becomes essential in classroom discourse.

To increase SSI inclusion in science teaching, teacher education programs need to incorporate SSI into their programs. Preservice teachers should not only learn about the SSI but also gain experience developing materials and lesson plans incorporating the SSI. To train preservice teachers to learn and use SSI effectively, teacher skills and the SSI curriculum need to be considered in detail. An SSI curriculum should include an interdisciplinary approach that aims to promote problem-solving, argumentation, perspective-taking skills, research skills, knowledge of NOS, and real-life problems and real data where students practice higher-order skills. The issue in science classrooms should also be controversial so students can explore it from different perspectives and improve their awareness of its moral and ethical dimensions (Kinskey, 2020). In addition, to support preservice teachers in successfully integrating SSI in their future classrooms, teacher educators should consider their character and values, implement perspective-taking activities, incorporate place-based education, and provide thorough SSI training in teacher education programs. This will help students develop their critical-thinking abilities and foster a deeper understanding of SSI.

8.2 SSI Context: Energy Demand in Muğla

With population growth and developing technology in the world, the demand for energy in the world has increased continuously (Masnadi et al., 2015). Today, a significant part of the energy need is obtained from fossil fuels. For example, in 2010, fossil fuels provided about 80% of all primary energy worldwide (Höök & Tang, 2013). The depletion of these energy sources, the increasing energy need, and the negative effects of fossil fuel consumption on the environment have caused people to seek new sources and increased the importance of renewable energy.

Similarly, despite the increase in energy production in Türkiye, dependence on energy exports has increased due to the increase in energy demand (Demirbaş & Bakis, 2004; Saraç & Bedir, 2014). In the years between 2000 and 2020, total energy production increased by 66.9%, but energy demand also increased by 85.3%. The ratio of total energy production to total energy demand decreased from 33% to 29.8% in the last 20 years (Yılmaz & Türkyılmaz, 2022). It can be predicted that the need for energy will rise in the near future due to new technological appliances and the population rise of Türkiye. Therefore, it becomes inevitable to encourage the public toward the use of domestic and renewable energy resources (Ünal, 2022; Yılmaz & Türkyılmaz, 2022).

According to the General Directorate of Renewable Energy, renewable energy is the type of energy obtained from cycles existing in nature, such as wind, geothermal, solar, biomass, hydraulic, and hydrogen energy. Non-renewable energy sources (e.g., oil, coal, and natural gas) have a limited lifespan and cause environmental pollution. To meet energy needs while protecting the natural environment, non-renewable energy sources should be replaced by renewable energy sources. Using renewable energy widely is essential for an economical and clean environment (Bozdoğan & Yiğit, 2014).

The use of renewable energy generation has increased in Türkiye (Oral, 2020). In the last 20 years between 2000 and 2020, Türkiye's total energy supply has increased by 85.3%, while oil has increased by 25.4%, coal by 77.3%, natural gas by 219.8%, hydraulics by 152.8%, geothermal, solar, and wind by 1386%, and the total of bio-energy, wood, garbage, and waste decreased by 52.2%. If the current policies in Türkiye are maintained, it is estimated that although the share of fossil fuels in the primary energy supply will decline from 83.4% to 68.6% in 2050, the amount consumed will not decrease in absolute terms and will increase by 3.75%. It is predicted that the share of renewable resources in the total energy supply will increase by 124% and the amount by 182%. The share of nuclear power plant investments will continue at 5% (Yılmaz & Türkyılmaz, 2022).

Muğla province is a region with high energy capacity due to its geographical location and geological structure. The province of Muğla has a variety of energy generation opportunities but cannot fully use its capacity. The potential to benefit from solar and wind energy in Muğla is above the Türkiye average. Similarly, although it has geothermal resources, these resources are not used in energy production. In terms of renewable energy in the region, hydroelectric energy is mostly produced. However, the main energy source in the region is three active thermal power plants that use the lignite deposits in the region. In light of this situation, the evaluation of renewable energy sources in meeting the energy needs in the energy studies planned to be carried out in Muğla province is of great importance in terms of protecting the environment by using a cleaner energy source (Özkan et al., 2016).

Muğla province is one of the leading cities in Türkiye in the field of tourism, and every year, especially in the summer months, the population and, therefore, the need for energy increase due to the tourists coming to the city. For this reason, energy production should be evaluated by considering both the increasing energy needs of Muğla and its potential to produce energy from renewable sources. In addition to these, Muğla province primarily makes a living from tourism, agriculture, forestry, and beekeeping. For this reason, it should not be ignored that the facilities to be established should cause the least damage to these occupational groups while investing in the resources that will meet the energy needs of Muğla. Although renewable energy sources are environmentally friendly, their negative effects are unexplored. For example, wind blades are a deadly mechanism for insects and bees in wind power plants. In addition to this lethal effect, the turbulence and pressure difference around the turbine also cause permanent physical damage to the bees (Gürbüz et al., 2021). Again, establishing wind power plants and solar power plants on fertile lands will not allow the agricultural use of the land. For this reason, it is of great importance to choose the location where these power plants will be established. Yenişehirlioğlu and Türkay (2021) examined a hotel in Vienna on sustainable energy management practices and current solutions in accommodation businesses. Part of this hotel is designed in a way that is called a passive house, which needs very little energy and can get all of this energy from renewable energy sources. The operator of the passive house stated that to operate with 100% renewable energy, 50% energy savings, and 50% renewable energy production should be done. The level of savings must be very high so that the needs can be met with renewable energy sources. However, hosting guests by meeting only their minimal needs contradicts the paradigm of accommodation businesses that support this luxury consumption. This type of consumption is among the main sources of income for accommodation businesses. However, this does not necessitate further integration of renewable energy. On the contrary, it draws attention to the issues to be considered when using renewable energy sources.

Learning about renewable energy is important both for the next generation to be informed about energy and to raise awareness about the conscious use of energy resources. For this purpose, the curriculum should include courses to learn about energy needs, energy efficiency, and environmental protection (Zafar et al., 2020). In addition, while planning the lessons in the curriculum, the limitation of energy resources and what needs to be done for a solution should be included (Marker, 1991; Keser, 2003). Using different perspectives while finding solutions to these problems also develops students' cognitive thinking skills (Kahn & Zeidler, 2016).

8.3 Implementation

In Türkiye, senior preservice teachers must complete their internships, which are called "teaching practice," in the last two semesters of the undergraduate program. In both semesters, their internship consists of two parts: the theoretical part and the application part. The application part takes place in public middle schools, and preservice teachers spend 6 h per week attending science courses under the supervision of a science teacher. The theoretical part takes place at the university for 2 h per week, and the course instructor goes over the teaching strategies, assessment strategies, lesson planning, and classroom management issues and discusses them based on preservice teachers' experiences in the middle school science classroom. The difference between the first and second semesters in terms of preservice teacher roles is that in the first part, preservice teachers observe and learn more about the school, while in the second part, preservice teachers are more focused on teaching individually, and they also develop lesson plans and activity plans individually.

To ensure preservice science teachers have effective experiences in teaching science courses, four students are enrolled in each section of a teaching practice course with a middle school teacher. The authors taught two sections of the course. In this study, out of thirty-seven, eight preservice science teachers (PSTs) registered in the Teaching Practice II course that the authors teach, and they voluntarily participated in this study. All the participants were female. Five students took a "Renewable Energy Sources" course earlier, and three took the course simultaneously with the study. Even though none of the students took a course on SSI, three asserted they learned about it in another course related to science teaching.

The course included SSI-infused place-based instruction with a specific focus on Muğla and the increasing energy demand in the region. The course content addressed a real-world issue, providing students with a meaningful learning experience. Furthermore, preservice science teachers' engagement with the local community was encouraged through the sharing of videos and news in the media relevant to Muğla. The Muğla Environmental Platform was also introduced to students, encouraging active participation in such communities. This matches the course goals of students, as renewable energy is included in the middle school science program, and SSI can be used as an effective teaching strategy. During implementation, preservice science teachers experienced and learned SSI-based instruction and developed lesson plans related to renewable energy. Thus, renewable energy resources were covered in detail within the scope of teaching SSI. The implementation lasted for 4 weeks. In general, the course is taught face-to-face. However, due to a significant earthquake in Türkiye, all of the courses in the university were shifted to an online format during the implementation period. Thus, this 4-week process took place on the Zoom portal, which the university recommends for the conduct of distance lessons. Zoom enabled students to participate in the course discussions actively and to share their materials easily by using the share screen features. Two instructors (authors of this chapter) taught the course, and during the lessons, both instructors actively participated in lectures and discussions.

The implementation is centered around the question, "What is the current state of Muğla's increasing energy demand, and what potential solutions can be implemented to meet this growing need?" We selected the issue based on Hancock et al.'s (2019) suggestion. We chose renewable energy sources because both instructors have expertise in physics education, and one instructor specifically teaches renewable energy sources. Therefore, both instructors were familiar with and knowledgeable about the topic. Geographically and geologically, Muğla is well-positioned to make greater use of renewable energies. Additionally, the increase in Muğla's population during the summer months indicates a rise in energy demand. Moreover, since the topic aligns with middle school science course learning objectives, preservice teachers can incorporate it into their lesson plans. We prepared a course on the utilization of renewable energy sources in Muğla after considering all of these factors.

Before the implementation began, preservice science teachers were informed about the research, and their consent was obtained. Following that, questions were given to learn the knowledge and views of the preservice science teachers on renewable energy and to determine their views on the necessity of teaching about renewable energy in their own lessons.

1. *What are the renewable energy sources you are familiar with? If you know, what are the advantages and disadvantages? Please fill in the table.*

Name of renewable energy source	*Advantages*	*Disadvantages*

2. *What role do you think the use of renewable energy sources will play in the future? Do you think it is enough to meet our energy needs?*
3. *What should be considered in the teaching of renewable energy? And what should be the purpose of such teaching?*

In the first week, we started talking about increasing energy demand in the world and in Muğla. We discuss possible solutions to this increased demand. The students mainly stated that using renewable energy could solve the problem. Then, renewable energy sources and their advantages and disadvantages were explained. At the end of the first week, preservice science teachers were asked to find an activity plan on renewable energy. The reason to ask students to find an activity instead of writing on their own is to make preservice science teachers familiar with the lesson plans and later show them how the available lesson plans are limited. Additionally, at the end of the first week, the preservice science teachers were asked to answer the following questions in order to review their own energy consumption habits.

1. *What is the amount of electrical energy consumed in your home? Answer using your home's latest electricity bill.*
2. *What are the electric tools in your home?*
3. *Can you name three of these tools that you frequently use in your house and indicate the number of watts they utilize?*
4. *What other purposes do you use energy besides home consumption? Please provide a summary of your usage over the last week. (For example, you can answer the needs of transportation, food, etc.)*
5. *Choose a lesson or activity plan related to renewable energy and share your chosen plan with us. (You can copy and paste into this file.)*
6. *Why did you choose this activity plan? What is your favorite part of this plan?*
7. *What can you do to improve this activity plan?*

In the second week, preservice science teachers engaged in discussions about their energy consumption habits and the growing energy needs of themselves and the world. Preservice teachers stated that after responding to the questions, their awareness of energy consumption increased. They also discussed the role of technology in how our habits have changed throughout the years and its effects on increasing our energy needs. Then, how the energy needs are met, and the role of renewable energy in meeting these needs was discussed. In addition, information about the energy resources and renewable energy in the world, in Türkiye and Muğla, was provided to preservice teachers. Reports on energy (e.g., https://www.mmo.org.tr/kitaplar/turkiyenin-enerji-gorunumu-2022) and websites were visited (https://enerji.gov.tr/bilgi-merkezi-enerji, https://temizenerji.org/, https://www.enerjiatlasi.com/). Students also visited the websites themselves, and classroom discussions were held.

At the end of the second week, preservice science teachers were asked to watch two short videos as homework that enabled them to look at the disadvantages of renewable energy from different perspectives. Then, they were asked to provide their opinions on how they foresee a method to be followed in meeting the future energy needs of Muğla based on what they learned in the course and the views in the videos. In addition, preservice science teachers were asked to write down the changes they planned to make to their activity plans.

In our last lesson, you learned about energy production, energy consumption, and the role of renewable energy in the energy production process in the world, Türkiye and Muğla. Now, please watch the videos at https://www.youtube.com/watch?v=RN0wM8m3lI0 and https://www.youtube.com/watch?v=EvnBJmR7gOI. Based on all this information,

1. *What kind of method do you foresee for meeting the future energy needs of Muğla?*
2. *What changes do you plan to make in your lesson plans? Explain with specific examples. (Please provide at least three examples.)*

In the third week, after discussing the previous week's questions, students asserted that the videos provided a new way of looking at renewable energy sources. Instructors also shared news on protests against renewable energy in Muğla. Instructors also informed students about the Muğla Environmental Platform, created by citizens and non-governmental organizations from all towns to protect nature and common living spaces in the Muğla region. They showed the website to the students. Instructors also informed students that the news and the website have a certain perspective. Following that, preservice science teachers discussed meeting Muğla's energy needs. They shared the increased awareness that they need to pay attention to circumstances before establishing a renewable energy plant. Following that, SSI-based teaching strategies were explained to the preservice science teachers, and detailed examples were provided to assist in how they should be integrated into the lesson. Finally, there was an in-class discussion with the preservice science teachers on how they could integrate them into their own plans. Even though SSI-based instruction was briefly explained to preservice science teachers before the implementation started, this was the first time it was explained how to incorporate their lessons in detail.

At the end of the third week, preservice science teachers were asked to make changes in their lesson plans by considering the SSI-based teaching method. In addition, in order to prepare for the lesson in the fourth week, different roles (farmer, university student, tourism operator, and beekeeper) were assigned to preservice science teachers, and they were asked to do research considering the roles given and what needs to be done to meet the energy needs in Muğla. Preservice teachers are encouraged to use Toulmin's argumentation model, and an example of it is shared.

Hello friends,

1. *Please share with us the changes you have made to your event plan. When sending your plans, mark the places where you made changes. And explain why you made such a change or addition (1–2 paragraphs are enough).*

In this week's lesson, we will discuss the effects of renewable energy from different perspectives. Please imagine yourself as a student, a farmer, a beekeeper, or a tourism operator (see who you are on the list) living in Muğla. It is discussed in Muğla city council that the energy needs are not met, and a new facility should be established for this. You are also expected to participate in the discussion about which power plant should be built. Next week, we will have our city council discussion in our class. We expect you to actively participate in this discussion with concrete examples and scientific evidence. For this purpose, answer the following questions, taking into account your role: Use the examples below when answering.

2. *Which plant or facilities should be established?*
3. *Will the proposed facility benefit you? (Support with scientific evidence and examples.) Justification and supporting statements should be used.*
4. *What might be the adverse effects of another facility on you? (Support with scientific evidence and examples. Also, add limiting and rebuttal.)*

In the fourth week, preservice science teachers were asked to examine renewable energy from different perspectives based on a scenario of "city council discussion about the energy need not being met in the city of Muğla and what can be done." Preservice science teachers acted as if they were farmers, beekeepers, tourism operators, or university students. They discussed how the renewable energy system would affect their lives. For example, a preservice science teacher, acting as a beekeeper, talked about how bees are important for the improvement of the city. Honey is important for local agriculture. She provided information on how the wind energy systems could affect the bees negatively. Another preservice science teacher, in a university student role, mentioned the high cost of living in Muğla as a student. She asserted that solar plants could be a logical solution for reducing electric bills, especially considering that Muğla is located in the southern part of the country and receives abundant sunlight. The tourism operator asserts that while installing solar panels, the land should be considered carefully as there are a lot of tourist and historical sites in the area. At the end of the city council meeting, preservice science teachers decided to put solar panels on the roofs of public buildings and try to reduce energy consumption. After the council meeting, preservice science teachers shared ideas from the meeting, which they could integrate into their own activity plans. Preservice science teachers were asked for the final version of their activity plans at the end of the last week.

8.4 Outcomes

Even though five of the students had already taken a course on renewable energy sources and three of them were taking a course on renewable energy sources simultaneously with this study, most of them had limited knowledge on the topic. The students asserted that they had some knowledge about renewable energy sources, but after completing this course, they learned more about renewable energy sources, specifically those in Muğla. One of the students (PST-5) stated, "*I didn't know we had that many renewable energy plants. This course helped me learn about them.*" Prior to the implementation, they were only focusing on the positive effects of renewable energy sources. Throughout the course, we showed a few videos highlighting the effects of renewable energy sources, and the students started paying attention to different aspects of renewable energy sources. All of the students mentioned that their awareness of evaluating circumstances increased after this course. PST-2 said, "*Prior to this course, I was all abroad with renewable energy. I believed that we should use renewable energy everywhere and as much as we can. After taking this course, I realized that everything is not so rosy. When setting up renewable energy power plants, we should research the location, its environmental impact, and the cost involved. I still support renewable energy, but my awareness has increased.*" Only one student mentioned that she became more aware that renewable energy is commonly used in different sectors. PST-6 said, "*With this course, I learned that renewable energy is used in many different sectors, such as tourism, and it affects these sectors.*"

The students expressed that prior to this, they relied on limited resources, such as textbooks and platforms like Twitter and Google, to acquire information about renewable energy sources when relevant information arose. However, they have now begun to expand their research by exploring various websites, reading news articles, and delving into scientific articles. These additional sources serve as valuable supplements to enhance their understanding of the topic. The students mentioned that using news articles and websites related to the environment provides them with different perspectives. PST-4 said, "*There are a lot of websites that provide similar information, short descriptions of the topic, and similar examples. News articles provide different perspectives. So after the course, I started using news articles and scientific articles.*"

The preservice science teachers mentioned that they enjoyed the fact that the issue discussed was related to Muğla because it allowed them to learn more about the city and have the opportunity to visit relevant places if needed. PST-1 said, "*Discussing local issues helps me incorporate local examples into future lessons. I am familiar with the city, and I have the opportunity to visit, see, and ask questions to the responsible persons at the facilities.*" PST-8 also mentioned that she added solar panels to her plans after learning about them in the course, and she went to the campus to gather more information. PST-4, PST-6, and PST-8 mentioned that discussing local issues helped them better understand the examples and also motivated them. Additionally, PST-7 and PST-8 stated that when local issues were integrated into the lessons, it would encourage students to actively engage in finding solutions to issues.

During the interviews, the preservice science teachers highlighted that they enjoyed role-playing and discussing the issues from different perspectives. Perspective-taking made them aware of different issues, especially when combined with the context of Muğla, allowing them to understand the topic better. The students mentioned that they enjoyed using place-based SSI because it enabled them to understand the issue better as they were familiar with the place and the culture. Some preservice science teachers mentioned that they could discuss the issue with their friends and families since it was affecting them, too. PST-5 mentioned that she was able to discuss the topic with a friend who works at the Yatağan thermal power plants. PST-7 discussed the topic with the father of a friend who is a beekeeper. This enabled them to see the effects of wind power on the family, which facilitated meaningful engagement. All students asserted that using different perspectives during classroom activities, such as role-playing, enabled them to discuss the issue from a holistic view. PST-3 said, *"Looking at the issue from different perspectives was guiding, as it enabled us to gain a comprehensive understanding of the topic."* PST-2 added, *"While I was supporting solar energy, my friend didn't. Their reasoning made a lot of sense, and it made me realize the importance of understanding people's varying needs and perspectives. This helped us see the issue from a broader and more comprehensive standpoint."* PST-2 also used perspective-taking in her internship classroom and asserted that using perspective-taking engaged middle school students and provided them with a holistic view to understand the topic.

The preservice science teachers' lesson plans also changed. Initially, they found lesson plans based on renewable energy that either focused on lecturing about renewable energy, showing videos, or conducting cookbook experiments related to renewable energy without any connection to the Muğla province and only focusing on the positive aspects of renewable energy. Most of the final lesson plans integrated argumentation and perspective-taking and provided local examples in their lessons. Some students added field trips and incorporated local issues. The students mentioned that their lesson plans evolved through classroom discussions about their plans. During these discussions, both instructors and peers provided feedback. During the interview, PST-3, PST-6, and PST-7 mentioned that they wanted to include field trips in their future lesson plans. Figure 8.1 is an example showcasing the evolution of the students' activity from their initial to final versions in the lesson plan:

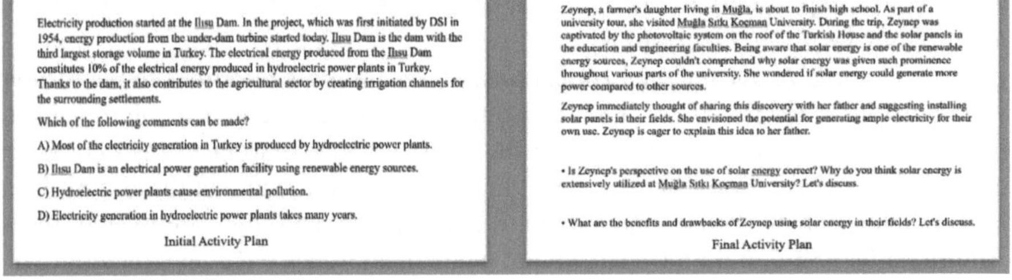

Electricity production started at the Ilısu Dam. In the project, which was first initiated by DSI in 1954, energy production from the under-dam turbine started today. Ilısu Dam is the dam with the third largest storage volume in Turkey. The electrical energy produced from the Ilısu Dam constitutes 10% of the electrical energy produced in hydroelectric power plants in Turkey. Thanks to the dam, it also contributes to the agricultural sector by creating irrigation channels for the surrounding settlements.

Which of the following comments can be made?

A) Most of the electricity generation in Turkey is produced by hydroelectric power plants.

B) Ilısu Dam is an electrical power generation facility using renewable energy sources.

C) Hydroelectric power plants cause environmental pollution.

D) Electricity generation in hydroelectric power plants takes many years.

Initial Activity Plan

Zeynep, a farmer's daughter living in Muğla, is about to finish high school. As part of a university tour, she visited Muğla Sıtkı Koçman University. During the trip, Zeynep was captivated by the photovoltaic system on the roof of the Turkish House and the solar panels in the education and engineering faculties. Being aware that solar energy is one of the renewable energy sources, Zeynep couldn't comprehend why solar energy was given such prominence throughout various parts of the university. She wondered if solar energy could generate more power compared to other sources.

Zeynep immediately thought of sharing this discovery with her father and suggesting installing solar panels in their fields. She envisioned the potential for generating ample electricity for their own use. Zeynep is eager to explain this idea to her father.

• Is Zeynep's perspective on the use of solar energy correct? Why do you think solar energy is extensively utilized at Muğla Sıtkı Koçman University? Let's discuss.

• What are the benefits and drawbacks of Zeynep using solar energy in their fields? Let's discuss.

Final Activity Plan

Fig. 8.1 Development of student activity

8.5 Transfer to the Global Context

Global energy demand is a pervasive challenge that requires immediate attention. While renewable energy presents a potential avenue to address this issue, finding comprehensive solutions remains a complex endeavor. In this chapter, we aim to explore the energy demand and potential solutions specific to the region of Muğla. Fortunately, Muğla offers a supportive environment for widely adopting renewable energy sources. Instructors should carefully select the issue based on geographical and geological considerations, as well as the needs of the local population. As Kim et al. (2020) highlighted, this approach will foster students' sense of attachment and engagement with the lesson. Nonetheless, considering the diverse energy needs across different regions worldwide would provide valuable insights and even more diverse perspectives, as the implementation and utilization of renewable energy systems depend on each area's unique characteristics and requirements.

Perspective-taking is pivotal in fostering effective argumentation and facilitating place-based SSI teaching methodologies. Specifically, within the context of place-based SSI, students actively understand and address local issues, thereby cultivating a holistic approach to problem-solving in their real-life experiences. This implementation fosters a stronger connection between students and real-world contexts. In our implementation, we provide opportunities for preservice science teachers to actively participate in place-based SSI and assist them while developing their lesson plans. By immersing themselves in the local context and experiencing the learning process firsthand, they receive support to elevate their teaching proficiency. Both instructors and peers offer guidance and assistance, aligning with existing literature suggesting that such support enhances content knowledge, pedagogical skills, and self-efficacy in teaching (Kinskey& Zeidler, 2021; Sibic & Topçu, 2020).

Furthermore, it is recommended that future research place greater emphasis on investigating classroom practices. Additionally, in-service teachers should be offered frequent training programs focusing on cultivating expertise in current topics, leveraging scientific research resources, and enhancing technological competencies. Additionally, enriching learning environments with diverse teaching materials to support SSI learning would be beneficial in promoting effective and engaging educational experiences (Aydın et al., 2021).

8.6 Conclusion

The current study addressed the pressing issue of Muğla's increasing energy demand and finding viable solutions. Through the integration of place-based SSI instruction, this research aimed to provide preservice science teachers with the opportunity to explore the topic of renewable energy within the specific context of Muğla. The study's objectives encompassed not only equipping preservice science teachers with knowledge about renewable energy but also fostering their ability to critically analyze and propose solutions to the identified problem while considering the unique circumstances of Muğla province. An important aspect of the investigation involved investigating the impact of perspective-taking on students' cognitive and affective engagement. Furthermore, students were required to develop their own lesson plans related to renewable energy. The lesson plans were then analyzed to determine the extent to which aspects of what they learned in class were integrated into their lesson plans.

Incorporating perspective-taking enabled students to develop a holistic perspective and gain awareness of how the problem impacts the people around them (Leung et al., 2015; Newton & Zeidler, 2020; Pouliot, 2008; Zeidler et al., 2005). Moreover, perspective-taking skills were found to enhance argumentation (Leung & Cheng, 2020; Kahn & Zeidler, 2016; Sadler et al., 2007). Following the recommendation of Choi and Lee (2021), various methods were employed, including role-playing, the use of newspaper articles, websites, YouTube videos, and research articles, to integrate perspective-taking into the instructional approach effectively. Additionally, students were encouraged to share their own experiences through engaging in classroom discussions and weekly questions.

The study's findings indicate that students developed a comprehensive understanding of the advantages and disadvantages of renewable energy sources. Furthermore, they gained a deeper knowledge of Muğla and its potential for renewable energy. Students expressed that the use of a local issue as the focal point of the study motivated them and enhanced their understanding and experiential learning. The exploration of different perspectives was particularly valued by students as it broadened their horizons. Overall, the feedback from students regarding the implementation was positive. They reported enjoying active engagement, conducting research, and learning about Muğla's energy needs and renewable energy opportunities. The role-playing activities, in particular, were well-received as they allowed students to adopt diverse perspectives. Students also highlighted the benefits of in-class discussions, which helped them improve their lesson plans. The supportive environment provided by instructors and peers, along with constructive feedback on their lesson plans, contributed to their positive experiences. However, students suggested that the implementation could have been further enhanced in a face-to-face course, as it would have facilitated more fruitful classroom discussions. It is worth noting that technical issues prevented some students from participating in these discussions.

Ultimately, this study significantly contributes to advancing effective instructional practices in promoting students' engagement with SSI within localized contexts and using perspective-taking skills. To further enhance our understanding, future research should explore the long-term effects of incorporating place-based SSI instruction, examining students' environmental awareness and decision-making over an extended period. Additionally, it is essential to delve into specific strategies such as role-playing scenarios, multimedia resources, real-life case studies, and pedagogical approaches that effectively foster perspective-taking and deepen students' engagement in SSI. Furthermore, gaining a comprehensive understanding of the benefits and challenges associated with implementing place-based SSI instruction in diverse learning environments, including both face-to-face and online, will contribute to optimizing instructional practices. Through these research directions, we can continue to refine and expand our knowledge of effective instructional methods that promote students' active involvement and holistic understanding of SSI in a localized context.

References

Apaydın, Z., & Kandemir, M. A. (2018). Sınıf öğretmenlerinin fen bilimleri dersinde kullandıkları öğretim yöntem, teknik ve değerlendirme araçlarına ilişkin görüşleri [Opinions of classroom teachers on teaching methods, techniques and assessment tools used in science lessons]. *Dicle Üniversitesi Ziya Gökalp Eğitim Fakültesi, 33*, 70–78. https://doi.org/10.14582/DUZGEF.1884

Aydın, G. Ş., Sarıbaş, D., Özalp, D., & Yılmaz, Ş. (2021). Biyoloji öğretmenlerinin sosyo-bilimsel konuların öğretimine yönelik görüşlerinin incelenmesi [Examination of biology teachers' views on teaching socio-scientific issues]. *Mersin Üniversitesi Eğitim Fakültesi Dergisi, 17*(1), 161–181. https://doi.org/10.17860/mersinefd.827736

Bang, M., & Marin, A. (2015). Nature–culture constructs in science learning: Human/non/human agency and intentionality. *Journal of Research in Science Teaching, 52*(4), 530–544. https://doi.org/10.1002/tea.21204

Barrett, S. E., & Nieswandt, M. (2010). Teaching about ethics through socioscientific issues in physics and chemistry: Teacher candidates' beliefs. *Journal of Research in Science Teaching, 47*, 380–401. https://doi.org/10.1002/tea.20343

Borgerding, L., & Dagistan, M. (2018). Preservice science teachers' concerns and approaches for teaching socioscientific and controversial issues. *Journal of Science Teacher Education, 29*(4), 283–306. https://doi.org/10.1080/1046560X.2018.1440860

Bozdoğan, A. E., & Yiğit, D. (2014). Öğretmen adaylarının alternatif enerji kaynaklarına yönelik görüşlerinin farklı değişkenler açısından incelenmesi [Examination of preservice teachers' views on alternative energy sources in terms of different variables]. *Electronic Journal of Education Sciences, 3*(6).

Brocos, P., & Jiménez-Aleixandre, M. P. (2020). What to eat here and now: Contextualization of scientific argumentation from a place-based perspective. In I. S. Tapia (Ed.), *International perspectives on the contextualization of science education* (pp. 15–46). Springer. https://doi.org/10.1007/978-3-030-27982-0_2

Choi, Y., & Lee, H. (2021). Exploring the effects of implementing a research-based SSI program on students' understanding of SSI and willingness to act. *Asia-Pacific Science Education, 7*(2), 477–499. https://doi.org/10.1163/23641177-bja10033

Demirbaş, A., & Bakis, R. (2004). Energy from renewable sources in Turkey: Status and future direction. *Energy Sources, 26*(5), 473–484. https://doi.org/10.1080/00908310490429759

Evren Yapıcıoğlu, A., & Kaptan, F. (2018). Sosyobilimsel durum temelli öğretim yaklaşımının argümantasyon becerilerinin gelişimine katkısı: Bir karma yöntem araştırması [Contribution of socioscientific situation-based teaching approach to the development of argumentation skills: A mixed method research]. *Ondokuz Mayıs Üniversitesi Eğitim Fakültesi Dergisi, 37*(1), 39–61.

Gürbüz, E. Y., Altıntaş, A., Sürücü, B., & Tuncer, A. D. (2021). Rüzgar türbinlerinin yaban hayatına etkilerinin incelenmesi [Investigation of the effects of wind turbines on wildlife]. *Politeknik Dergisi, 1*(1), 953–962. https://doi.org/10.2339/politeknik.741965

Hancock, T. S., Friedrichsen, P. J., Kinslow, A. T., & Sadler, T. D. (2019). Selecting socioscientific issues for teaching: A grounded theory study of how science teachers collaboratively design SSI-based curricula. *Science & Education, 28*, 639–667. https://doi.org/10.1007/s11191-019-00065-x

Han-Tosunoğlu, Ç., & İrez, S. (2017). Biyoloji öğretmenlerinin sosyobilimsel konularla ile ilgili anlayışları [Biology teachers' understanding of socioscientific issues]. *Uludağ Üniversitesi Eğitim Fakültesi Dergisi, 30*(2), 833–860. https://doi.org/10.19171/uefad.369244

Herman, B. C. (2015). The influence of global warming science views and sociocultural factors on willingness to mitigate global warming. *Science Education, 1*(1), 1–38. https://doi.org/10.1002/sce.21136

Herman, B. C. (2018). Students' environmental NOS views, compassion, intent, and action: Impact of place-based socioscientific issues instruction. *Journal of Research in Science Teaching, 55*(4), 1–29. https://doi.org/10.1002/tea.21433

Herman, B. C., Sadler, T. D., Zeidler, D. L., & Newton, M. H. (2018). A socioscientific issues approach to environmental education. In G. Reis & J. Scott (Eds.), *International perspectives on the theory and practice of environmental education: A reader* (Environmental discourses in science education) (Vol. 3). Springer. https://doi.org/10.1007/978-3-319-67732-3_11

Herman, B. C., Zeidler, D. L., & Newton, M. (2020). Students' emotive reasoning through place-based environmental socioscientific issues. *Research in Science Education, 50*, 2081–2109. https://doi.org/10.1007/s11165-018-9764-1

Herman, B. C., Newton, M. H., & Zeider, D. (2021). Impact of place-based socioscientifc issues instruction on students' contextualization of socioscientifc orientations. *Science Education, 105*(4), 585–627. https://doi.org/10.1002/sce.21618

Herman, B. C., Poor, S. V., Oertli, R. T., & Schulte, K. (2022). Promoting young learners' NOS views through place-based SSI instruction. *Science & Education, 1–46*, 947–992. https://doi.org/10.1007/s11191-022-00353-z

Höök, M., & Tang, X. (2013). Depletion of fossil fuels and anthropogenic climate change – A review. *Energy Policy, 52*, 797–809. https://doi.org/10.1016/j.enpol.2012.10.046

Kahn, S., & Zeidler, D. L. (2016). Using our heads and HARTSS*: Developing perspective-taking skills for socioscientific reasoning (*Humanities, ARTS, and Social Sciences). *Journal of Science Teacher Education, 27*(3), 261–281. https://doi.org/10.1007/s10972-016-9458-3

Kahn, S., & Zeidler, D. L. (2019). A conceptual analysis of perspective taking: Positioning a tangled construct within science education and beyond. *Science & Education, 28*(6–7), 605–638. https://doi.org/10.1007/s11191-019-00044-2

Keser, Ö. (2003). Energy, environment, and education relationship, in developing countries' policies: A case study for Turkey. *Energy Sources, 25*(2), 123–133. https://doi.org/10.1080/00908310390142181

Kim, G., Ko, Y., & Lee, H. (2020). The effects of community-based socioscientific issues program (SSICOMM) on promoting students' sense of place and character as citizens. *International Journal of Science and Mathematics Education, 17*(4), 1–20. https://doi.org/10.1007/s10763-019-09976-1

Kinskey, M. (2020). *Developing and facilitating socioscientific issues based science lessons: Elementary preservice teachers' experiences* (Publication no. 27743561). Soctoral dissertation, University of South Florida. ProQuest Dissertations Publishing.

Kinskey, M., & Zeidler, D. (2021). Elementary preservice teachers' challenges in designing and implementing socioscientific issues-based lessons. *Journal of Science Teacher Education, 32*(3), 350–372. https://doi.org/10.1080/1046560X.2020.1826079

Kurdyavtsev, A., Stedman, R. C., & Krasny, M. E. (2012). Sense of place in environmental education. *Environmental Education Research, 18*(2), 229–250. https://doi.org/10.1080/13504622.2011.609615

Kutluca, A. Y., & Aydin, A. (2017). Changes in pre-service science teachers' understandings after being involved in explicit nature of science and socioscientific argumentation processes. *Science & Education, 26*, 637–668. https://doi.org/10.1007/s11191-017-9919-x

Lee, H., & Yang, J. (2019). Science teachers taking their first steps toward teaching socioscientific issues through collaborative action research. *Research in Science Education, 49*(1), 51–71. https://doi.org/10.1007/s11165-017-9614-6

Lee, H., Abd-El-Khalick, F., & Choi, K. (2006). Korean science teachers' perceptions of the introduction of socioscientific issues into the science curriculum. *Canadian Journal of Science, Mathematics and Technology Education, 6*(2), 97–117. https://doi.org/10.1080/14926150609556691

Lee, H., Chang, H., Choi, K., Kim, S. W., & Zeidler, D. L. (2012). Developing character and values for global citizens: Analysis of pre-service science teachers' moral reasoning on socioscientific issues. *International Journal of Science Education, 34*(6), 925–953. https://doi.org/10.1080/09500693.2011.625505

Lee, H., Kyunghee, C., Kim, S., Jungsook, Y., Krajcik, J. S., Herman, B. C., & Zeidler, D. L. (2013). Socioscientific issues as a vehicle for promoting character and values as global citizens. *International Journal of Science Education, 35*(12), 2079–2113. https://doi.org/10.1080/09500693.2012.749546

Leung, J. S. C., & Cheng, M. M. W. (2020). Conceptual change in socioscientific issues: Learning about obesity. *International Journal of Science Education, 42*(18), 3134–3158. https://doi.org/10.1080/09500693.2020.1856966

Leung, J. S. C., Wong, A. S. L., & Yung, B. H. W. (2015). Understandings of nature of science and multiple perspective evaluation of science news by non-science majors. *Science & Education, 24*(7–8), 887–912. https://doi.org/10.1007/s11191-014-9736-4

Marker, G. W. (1991). Has the time come to revive energy education? *The Social Studies, 82*(5), 183–187. https://doi.org/10.1080/00377996.1991.9958333

Masnadi, M. S., Grace, J. R., Bi, X. T., Lim, C. J., & Ellis, N. (2015). From fossil fuels towards renewables: Inhibitory and catalytic effects on carbon thermochemical conversion during co-gasification of biomass with fossil fuels. *Applied Energy, 140*, 196–209. https://doi.org/10.1016/j.apenergy.2014.12.006

Mayer, S. F., Frantz, C. M., Bruehlman-Senecal, E., & Dolliver, K. (2009). Why is nature beneficial? The role of connectedness to nature. *Environment and Behavior, 41*(5), 607–643. https://doi.org/10.1177/0013916508319745

Ministry of National Education. (2013). *İlköğretim kurumları (ilkokullar ve ortaokullar) fen bilimleri dersi (3, 4, 5, 6, 7 ve 8. sınıflar) öğretim programı* [Primary education institutions (primary and secondary schools) science lesson (3, 4, 5, 6, 7 and 8th grades) curriculum]. T.C Millî Eğitim Bakanlığı Talim ve Terbiye Kurulu Başkanlığı.

Nargund-Joshi, V. (2020). Reflections from a science teacher educator: Supporting pre-service teachers to teach science in a contextualized manner. In I. S. Tapia (Ed.), *International perspectives on the contextualization of science education* (pp. 147–157). Springer. https://doi.org/10.1007/978-3-030-27982-0_7

Newton, M. H., & Zeidler, D. L. (2020). Developing socioscientific perspective taking. *International Journal of Science Education, 42*(8), 1302–1319. https://doi.org/10.1080/09500693.2020.1756515

Okur Akçay, N., Akçay, A., & Kurt, M. (2016). Ortaokul öğretmenlerinin öğretim yöntem ve tekniklerine yönelik görüş ve yeterliklerinin incelenmesi [Examining the opinions and competencies of secondary school teachers about teaching methods and techniques]. *Eğitim ve Öğretim Araştırmaları Dergisi, 5*(1), 333–342.

Oral, M. (2020). Yenilenebilir enerji kaynaklarına ilişkin bir farkındalık araştırması [An awareness survey on renewable energy sources]. *Türkiye Bilimsel Araştırmalar Dergisi, 5*(2), 286–296.

Özkan, R., Gökdemir, A., & Şen, F. (2016). Muğla bölgesinin enerji kaynakları ve Muğla'da mevcut enerji üretimi [Energy resources of Muğla region and current energy production in Muğla]. *Küresel Mühendislik Çalışmaları Dergisi, 3*(1), 19–31.

Pouliot, C. (2008). Students' inventory of social actors concerned by the controversy surrounding cellular telephones: A case study. *Science Education, 92*(3), 543–559. https://doi.org/10.1002/sce.20274

Presley, M. L., Sickel, A. J., Muslu, N., Merle-Johnson, D., Witzig, S. B., Izci, K., & Sadler, T. D. (2013). A framework for socio-scientifc issues based education. *Science Educator, 22*(1), 26–32.

Reis, P., & Galvao, C. (2009). Teaching controversial socio-scientific issues in biology and geology classes: A case study. *Electronic Journal of Science Education, 13*(1), 162–185.
Sadler, T. D. (2004). Informal reasoning regarding socioscientific issues: A critical review of research. *Journal of Research in Science Teaching, 41*(5), 513–536. https://doi.org/10.1002/tea.20009
Sadler, T. D. (2009). Situated learning in science education: Socio-scientifc issues as contexts for practice. *Studies in Science Education, 45*(1), 1–42. https://doi.org/10.1080/03057260802681839
Sadler, T. D., & Zeidler, D. L. (2005). Patterns of informal reasoning in the context of socioscientific decision making. *Journal of Research in Science Teaching, 42*(1), 112–138. https://doi.org/10.1002/tea.20042
Sadler, T. D., Barab, S. A., & Scott, B. (2007). What do students gain by engaging in socioscientific inquiry? *Research in Science Education, 37*, 371–391. https://doi.org/10.1007/s11165-006-9030-9
Saraç, E., & Bedir, H. (2014). Sınıf öğretmenlerinin yenilenebilir enerji kaynakları ile ilgili algıları üzerine nitel bir çalışma [A qualitative study on primary school teachers' perceptions of renewable energy sources]. *Kara Harp Okulu Bilim Dergisi, 24*(1), 19–45.
Saunders, K. J., & Rennie, L. J. (2013). A pedagogical model for ethical inquiry into socioscientifc issues in science. *Research in Science Education, 43*(1), 253–274. https://doi.org/10.1007/s11165-011-9248-z
Semken, S., & Freeman, B. C. (2008). Sense of place in the practice and assessment of place-based science teaching. *Science Education, 92*, 1042–1057. https://doi.org/10.1002/sce.20279
Sibic, O., & Topcu, M. S. (2020). Pre-service science teachers' views towards socio-scientific issues and socio-scientific issue-based instruction. *Journal of Education in Science. Environment and Health (JESEH), 6*(4), 268–281. https://doi.org/10.21891/jeseh.749847
Ünal, C. (2022). *Türkiye'de enerji sektörünün genişletilmiş ekserji analizi kullanılarak değerlendirilmesi* [Evaluation of the energy sector in Turkey using extended exergy analysis]. Unpublished master thesis. Bilecik Şeyh Edebali Üniversitesi.
Yenişehirlioğlu, E., & Türkay, O. (2021). Yenilenebilir enerji teknolojilerinin konaklama işletmelerinde kullanımı ve turizm arzındaki paradigma kırılması: Avusturya Stadhale Hotel incelemesi [Use of renewable energy technologies in hospitality businesses and breaking the Paradigm in tourism supply: Austrian Stadhale Hotel review]. *Türk Turizm Araştırmaları Dergisi, 5*(1), 623–641. https://doi.org/10.26677/TR1010.2021.684
Yılmaz, Ş., & Türkyılmaz, O. (2022). *Dünyada ve Türkiye'de birincil enerji arzı.* TMMOB Makina Mühendisleri Odası [Primary energy supply in the world and in Turkey. TMMOB chamber of mechanical engineers]. Oda raporu. https://www.mmo.org.tr/kitaplar/turkiyenin-enerji-gorunumu-2022
Zafar, M. W., Shahbaz, M., Sinha, A., Sengupta, T., & Qin, Q. (2020). How renewable energy consumption contribute to environmental quality? The role of education in OECD countries. *Journal of Cleaner Production, 268*, 122–149. https://doi.org/10.1016/j.jclepro.2020.122149
Zeidler, D. L. (2014). Socioscientific issues as a curriculum emphasis: Theory, research and practice. In N. G. Lederman & S. K. Abell (Eds.), *Handbook of research on science education* (Vol. II, pp. 697–726). Routledge. https://doi.org/10.4324/9780203097267.ch34
Zeidler, D. L., Sadler, T. L., Simmons, M. L., & Howes, E. V. (2005). Beyond STS: A research-based framework for socioscientific issues education. *Science Education, 89*(3), 357–377. https://doi.org/10.1002/sce.20048

Chapter 9
Enhancing Preservice Teachers' Understanding of the Nature of Science through Place-Based Socioscientific Issue Instruction: The Sea-Filled Airport Case

Banu Avşar Erümit and Tuğba Yüksel

9.1 Introduction

In an era of heightened global environmental consciousness, it is critical to integrate an understanding of SSI into our educational endeavors. SSIs are often complex, involving scientific facts and societal, cultural, economic, moral, and ethical considerations (Sadler, 2004; Zeidler & Nichols, 2009). Place-based learning allows students to examine SSIs firsthand, explore these connections within their local environment, and enhance their understanding of them from multiple perspectives (Herman et al., 2021).

Place-based learning combines academic learning with real-world experiences, enabling students to connect with their local environment and community, thus fostering a sense of place while enhancing their understanding of academic concepts (Semken et al., 2017). The main advantage of place-based education is its ability to tailor to the distinct attributes of specific locations, thereby bridging the gap between academic settings and students' real-life experiences that often exist in numerous classrooms. A comprehensive knowledge of one's immediate environment can boost personal confidence and ability, improving the prospects for both their lives and future generations (Smith, 2002).

In this context, the sea-filled airport, a monumental infrastructure project built on reclaimed land from the sea, serves as an ideal platform for students to explore the dynamic intersection of science, society, technology, and the environment. While advantageous to the region's economy and tourism, this project is controversial due

B. Avşar Erümit (✉) T. Yüksel
Department of Mathematics and Science Education, Faculty of Education, Recep Tayyip Erdoğan University, Rize, Türkiye
e-mail: banu.avsar@erdogan.edu.tr; tugba.yuksel@erdogan.edu.tr

B. Namdar, E. Karahan (eds.), *Socioscientific Issues Focused Teacher Education*, Contemporary Trends and Issues in Science Education 60,
https://doi.org/10.1007/978-3-031-55233-5_9

to its potential impact on the sea ecosystem, surroundings, aesthetics, and local values (Erdoğan & Ercoşkun, 2021).

The sea-filled airport provides a tangible example of how advancements in engineering and aviation intersect with environmental concerns. This real-world scenario offers a valuable context for students to explore how such projects strive to mitigate their impacts on local ecosystems. To facilitate this understanding, we led preservice teachers on an immersive visit to the airport, allowing them to engage with the community and interact with individuals involved in the project. In this study, students sought answers about the challenges and solutions associated with constructing and operating a sea-filled airport, a controversial project due to its potential ecological implications. This direct experience gave them first-hand insights into the complexities and controversies surrounding the airport's construction and operation, granting a nuanced understanding of the issue.

Throughout the study, preservice teachers had the opportunity to engage with airport workers, an environmental specialist, and the airport's director. They were able to ask questions and gather first-hand insights about the measures taken to mitigate the ecological impacts of the airport. This active engagement equipped the participants with an understanding of the multifaceted nature of the issue, the decision-making processes in science, and the broad societal implications of such decisions.

Understanding the nature of science [NOS] is a fundamental aspect of preservice teachers' education and an essential component of scientific literacy. This understanding equips them with the capacity to navigate the multitude of SSIs they will confront in their future classrooms. In this study, we delve into the potential of place-based learning to enhance preservice teachers' grasp of the NOS. We specifically scrutinize the complexities inherent in an SSI: constructing and operating of the newly constructed sea-filled airport in northeastern Türkiye.

The sea-filled airport case underscores the societal and ethical implications of scientific decisions, presenting a concrete context for preservice teachers to engage in investigative processes. These prospective educators can examine how scientists, policymakers, and community members balance the potential ecological impacts, economic benefits, and societal needs associated with the airport. Furthermore, they can witness first-hand how engineers apply scientific principles creatively to design and implement strategies to minimize the environmental repercussions of the airport's construction and operational processes.

By appreciating how empirical evidence informs the selection of appropriate land for airport construction and the mitigation of potential environmental impacts, students gain insights into the role of empirical evidence in scientific decision-making and how societal needs and values shape scientific advancements (Sadler et al., 2004; Zeidler et al., 2002). This immersive experience could help refine their perspectives on the NOS.

This chapter focuses on place-based learning, an experiential pedagogy that combines academic knowledge with real-world experiences, to educate preservice social studies teachers about the complex interplay between scientific progress and societal impacts. The context for this exploration is the sea-filled airport, named Rize-Artvin Airport, an embodiment of this intricate relationship. The study investigates whether this place-based learning approach can improve preservice teachers' grasp of the NOS and socioscientific reasoning. By understanding how scientists make decisions and balance the trade-offs of such complex issues, preservice teachers are better prepared to instill a nuanced understanding of science within a societal context in their future students. Consequently, this study seeks to provide insights into the pedagogical benefits of place-based learning in enhancing pre-service teachers' NOS views in the context of real-world SSI.

9.1.1 Place-Based Socioscientific Issues in Promoting Preservice Teachers' Nature of Science Understanding

Science educators often face the challenge of providing relevant instructional contexts for students to develop meaningful learning. Numerous researchers have suggested that SSIs can serve as contextual entry points for engaging with the NOS (e.g., Bell et al., 2011; Eastwood et al., 2012; Khishfe, 2022; Khishfe et al., 2017; Leung, 2022). Previous research has shown that such knowledge fosters more informed and responsible decision-making and impacts students' decision-making factors within SSI contexts (Herman, 2015; Khishfe, 2012; Matkins & Bell, 2007).

Using SSIs as a context is a common method that likely enhances students' understanding of the NOS. However, as noted by Herman et al. (2021), the contextual elements of SSI interventions vary in their ability to provide substantial, high-quality pedagogical support and utilize students' connections with places, people, and environments affected by SSIs, with a spectrum that ranges from impoverished to rich. Historically, SSIs have been investigated predominantly in classrooms and through written scenarios and assessments and have been, at best, only tangentially related to environmental issues (Herman et al., 2019). Consequently, the current literature demands more emphasis on fully immersing students in authentic SSI experiences and greater integration between SSI and NOS contexts (Herman, 2018). One potential strategy that could cultivate this connection is a place-based approach.

A limited number of studies examined the impact of place-based SSI on enhancing students' NOS views. Herman (2018) found that place-based SSI instruction, concentrating on the scientific and sociocultural aspects of wolf reintroduction in Yellowstone, influenced the precision and contextual understanding of secondary students' views on the NOS. The study also found that the students' pro-environmental behavior was connected to their views on NOS, empathy for individuals grappling with contentious environmental issues, and pro-environmental intentions. Herman et al. (2019) found that students' explanations of the trophic cascade significantly improved in sophistication through place-based SSI instruction centered on the same SSI, the reintroduction of wolves in Yellowstone. Their results also showed a positive correlation between the complexity of students' post-instruction explanations of the trophic cascade and their understanding of the NOS.

In their recent research, Herman et al. (2023) explored the impact of the Missouri River SSI Education Program (MRSIP), a place-based educational initiative employing inquiry and argumentation strategies in both after-school environments and field settings. They aimed to determine if it could foster a deeper and more genuine understanding of the NOS among fourth-grade students. The study revealed that students involved in the MRSIP demonstrated considerably more advanced and non-stereotypical perspectives on how scientists conduct research and interpret the Missouri River SSI.

9.2 Sea-Filled Airports as Socioscientific Issues

There has been a growing trend among coastal countries to construct airports by leveraging marine resources to address land-based limitations and constraints (Yan, Wang, Yu, et al., 2013a; Zhao et al., 2010). Sea-filled airports provide additional space for infrastructure in areas where land is scarce or expensive. Notable examples include Kansai International Airport in Osaka, Japan, and Hong Kong International Airport. The development of sea airports has gained global attention within the aviation industry due to their potential to conserve land resources, reduce urban pollution, and enhance the safety of takeoff and landing operations (Zhao et al., 2010).

There is more to airport construction than finding a flat area. Suitable surrounding topographic conditions and local weather patterns conducive to aircraft takeoffs and landings are equally critical. Beyond physical conditions, practical considerations, such as proximity to city centers, integration with other transportation systems, and economic potential, are also critical in decision-making (Özür, 2018). Land reclamation for sea airports often requires extensive earthwork, which can dramatically alter geographic patterns and disrupt the natural equilibrium of the surrounding marine environment. This disruption can significantly harm the marine environment and coastal ecosystems (Zhao et al., 2010).

Whether inland or sea-filled airports, their harmful environmental impact is the biggest argument against their operations. Multiple sources contribute to emissions originating from airport activities, including aircraft, stationary sources, ground support equipment, ground access vehicles, and auxiliary power units. Among these sources, aircraft are the primary contributors to airport emissions, as stated in the Transportation Research Board Report (TRB, 2021). However, it is essential to acknowledge that air travel can also contribute to emissions from other sources. Surface access to airports, primarily through road vehicles traveling to and from airports, can result in additional emissions. Additionally, activities at airports, such as the heating and lighting of infrastructure and the operation of ground support equipment, can also contribute to emissions (TRB, 2021).

Global aviation activities significantly impact anthropogenic climate change by triggering a complex series of processes that result in a net warming of the Earth's surface. Key contributors to this effect include emissions from aviation, such as carbon dioxide (CO_2), nitrogen oxides (NOx), water vapor, soot, sulfate aerosols, and the formation of contrails, which lead to heightened cloudiness (Lee et al., 2021). Emissions from aircraft near or on the ground also pose a concern because, like other industrial facilities, they tend to be concentrated in specific areas such as airports, leading to the potential degradation of local air quality over time, as indicated by the National Research Council (NRC, 2002).

Although scientific and technological advancements in turbine engine technology have significantly reduced the noise and emissions produced by individual airplanes, the overall environmental consequences of aviation have become more evident due to the continued growth and demand for air travel (NRC, 2002). Aviation has experienced significant growth over the past few decades, with a substantial increase in revenue passenger kilometers from 109 billion kilometers per year to 8269 billion kilometers per year. As a result, there has been a notable increase in global aviation fuel consumption and CO_2 emissions, particularly in Asia and other developing regions (Lee et al., 2021). This growth can be attributed to the rapid expansion of civil aviation in these areas.

Noise pollution is a significant issue in air transportation. Both direct and indirect noise from aircraft can adversely affect humans and animals. Health effects on humans include problems with the cardiovascular system and sleep disorders (TRB, 2021). Sea-filled airports often offer noise pollution advantages over expanded inland airports (Yan, Wang, Wei, & Fu, 2013b). However, sea-filled airports also come with their own adverse impacts. These include increased sea pollution, disruption of marine ecosystems, alteration of natural habitats for aquatic organisms, negative impacts on fishing industries, harm to seabirds, and changes to the topography (Erdoğan & Ercoşkun, 2021).

Maintaining sea airports is a challenging task, equal to that of their construction. When constructing airports in areas with poor foundation geology and low stabilization, such as offshore embankments or reclaimed land with a short history, carefully considering the foundation layout is crucial. Sea-filled airports present a potential risk for land deformation and, as a result, may require more frequent maintenance and monitoring with various engineering measures (Bayik & Abdikan, 2021; Kadioglu, 2018). The challenges associated with these areas lie in the potential for cracks or collapses in buildings and infrastructure situated on such ground (Kadioglu, 2018).

Given the unstable nature of the foundation, it is essential to implement appropriate engineering solutions to ensure the stability and safety of the airport infrastructure. This may involve soil improvement techniques, such as compaction or stabilization, to enhance the foundation's load-bearing capacity and mitigate potential settlement issues. Additionally, advanced monitoring systems can be implemented to detect any ground movements or changes that may affect the integrity of the buildings and infrastructure (Bayik & Abdikan, 2021). For instance, Türkiye's relatively new first sea-filled airport, Giresun-Ordu Airport, has already experienced some deformation at the breakwater and the nearby taxiway (Bayik & Abdikan, 2021; Kadioglu, 2018).

9.3 Place-Based SSI Context: Rize-Artvin Airport

Due to the high number of rainy days in the Black Sea region, beach tourism has not developed significantly. However, the region possesses rich potential for highland and mountain tourism. Indeed, the Eastern Black Sea Mountains represent the most suitable area in our country for this type of tourism. Several reasons for this include the presence of numerous plateaus located close to the coast, each other, and residential areas; the rich vegetation of these plateaus, which is conducive to green tourism; the region's suitability for climatic tourism and mountaineering; and the availability of hot springs for thermal tourism. Rize, Artvin, and neighboring cities such as Trabzon serve as stopovers for Black Sea tours. Local and foreign tourists frequent the region during the spring and summer. The number of tourists has recently increased (Bekdemir & Özdemir, 2011). Consequently, with the future opening of the airport, the number of tourists visiting the region is expected to rise.

Rize-Artvin Airport, the second airport in Türkiye to be constructed on reclaimed sealand, is situated between the northeastern coastal cities of Rize and Artvin. Construction began in 2017, and flight operations were scheduled to begin in 2022. According to data revealed by the airport's director, supplied by the Ministry of Transport and Infrastructure of Türkiye, the airport spans an extensive area of 2.8 million m^2.

The airport's infrastructure includes a runway measuring 3 km in length and 45 m in width; three taxiways, each 265 m long and 45 m wide; and two aprons, with dimensions of 300 × 120 m and 120 × 120 m, respectively. The terminal building occupies an area of 32,000 m^2, with other enclosed spaces totaling 15,000 m^2. Notably, 100 million tons of rocks were transported from villages in and around Rize to construct the reclaimed land. The process of transporting rocks to the construction site spanned several months. According to the airport's director, each rock is interconnected like puzzle pieces; hence, to relocate one rock, all rocks must be displaced. This additional task, along with the transportation of rocks via trucks from miles away and their arrangement, has led to environmental concerns. These concerns are in addition to the environmental impacts of other processes involved in airport construction and operations.

9.4 Implementation

The study was implemented in the second author's out-of-school learning environment class. This course is offered to all academic departments within the faculty of education, including Turkish Language Teaching, Social Studies Teaching, Primary

Mathematics Teaching, Fine Arts Teaching, Classroom Teaching, and Guidance and Psychological Counseling. Enrolled students participated in a 15-week, face-to-face instructional session spanning 2 hours per week. The course comprehensively addressed theoretical and practical aspects associated with the out-of-school learning process. Before enrolling in this class, the preservice teachers had not undergone any prior coursework specifically addressing the nature of science and SSI. This study enlisted the participation of 24 pre-service teachers, comprising 20 females and four males. It is important to note, however, that a variance in participant attendance during the data collection days and the scheduled trip led to numerical disparities in the cohort during the data acquisition phase. Despite the existing class size, which comprised 32 students, the cohort engaged in this study consisted of 24 participants. Furthermore, in the scrutiny of both essay responses and VNOS-B data, analysis was confined to the subset of students who successfully completed both assessments. In this regard, essay data were analyzed for 24 students, whereas VNOS-B data were evaluated for 22 students.

The implementation of this study was carried out after the midterm exam. At the beginning and end of the intervention, teacher candidates were asked to complete the Views of Nature of Science (VNOS-B) questionnaire developed by Lederman et al. (2002). The Nature of Science assessment has evolved from predominantly forced-choice, quantitative methods in the late 1990s to a growing reliance on open-ended, qualitative assessments in the past two decades, with the Views of Nature of Science (VNOS) questionnaire being the most widely used tool since the early 2000s (Abd-El-Khalick et al., 2023). The VNOS-B instrument comprises seven open-ended questions designed to elucidate various aspects of their understanding of the nature of science. The components encompass (1) the tentativeness of science, (2) its foundation on empirical evidence, (3) the separation between observation and inference in data collection and analysis, (4) the subjectivity in data interpretation, (5) the necessity for creativity in the design and execution of scientific research, (6) the influence of social and cultural perspectives on interpretations, and (7) the distinction and connection between scientific theories and laws (Lederman, 2007).

The primary objective of employing VNOS-B was to ascertain if there were alterations in pre-service teachers' perspectives regarding the nature of science subsequent to their participation in a place-based SSI activity. SSI contexts offer excellent opportunities for cultivating accurate conceptions of the NOS. Complex, real-world problems like global warming are ideal for explicit NOS instruction. They present actual challenges that necessitate the use of evidence, data interpretation, acknowledgment of diverse viewpoints, and developing resolutions (Karisan & Zeidler, 2017). Due to the intensity of the course schedule and the fact that our pre-service teachers were not majoring in science, we did not focus on teaching science content or conduct other NOS-related activities. Instead, we introduced NOS concepts following a place-based visit, where we engaged them in explicit discussions and reflections on various scientific components. For example, we examined how empirical evidence is utilized by scientists in constructing and operating airports and the subjectivity of scientists regarding the environmental impact of sea-filled airports. However, certain aspects, like the distinction between theory and law, which were not directly observable during the place-based visit, were not explicitly addressed in this study. This approach aimed to provide a practical understanding of NOS in the context of real-world applications, especially tailored to accommodate the educational needs of non-science majors.

In addition to evaluating their views on the Nature of Science, we aimed to understand any changes in their decision-making mechanisms before and after the place-based SSI activity. To clarify the nature and extent of these changes, they were asked to express their decision-making with their reasoning in a concise essay format. The specific focus of the essay was to elucidate their decision regarding the construction or non-construction of the Rize-Artvin airport within the designated region.

Table 9.1, provided below, outlines the weekly implementation of the place-based SSI intervention.

Prior to implementing the place-based intervention, pre-service teachers were tasked with conducting comprehensive research about the construction phases of the Rize-Artvin airport construction within the designated location. This investigation encompassed the process of land reclamation from the sea alongside an examination of the perspectives held by various stakeholders, including local residents, political figures, environmental engineers, and other stakeholders, accompanied by a scrutiny of the underlying rationales informing their positions on the issue.

Before the place-based intervention, pre-service teachers were instructed to formulate a set of ten questions, each directed toward the airport authorities. These questions were designed to either corroborate the information acquired through the pre-service teachers' antecedent research or to elicit details that remained elusive during the course of their investigative endeavors. We encouraged them to note any information that supported their arguments during the visit. We first met with airport security and the environmental specialist at the airport. Accompanied by these officials, we toured various airport departments where the preservice teachers engaged in spontaneous conversations and asked their prepared questions. We visited the fire department, aprons, terminal buildings, airport control tower, and the director's office.

Table 9.1 The place-based SSI intervention

Week 8	Collecting pre-service teachers' essays and VNOS-B responses
Week 9	Pre-service teachers are tasked with engaging in research pertaining to the Rize-Artvin airport, with a concurrent examination of SSI involving a comprehensive discussion on their multifaceted nature and inherent complexity.
Week 10	Place-based implementation – Field trip to Rize-Artvin airport, during which pre-service teachers converse with an environmental specialist, workers, the airport's director, and passengers and ask them questions they prepared beforehand.
Week 11	Sharing examples of different sea-filled airports worldwide. Argumentation groups research and present their findings on the Rize-Artvin airport based on economy, politics, tourism, environment, science, and public perspectives.
Week 12	Collecting pre-service teachers' essays and VNOS-B responses

Preservice teachers asked questions about various aspects of the airport, including the economy, policy, tourism, environment, and public. The environmental specialist provided detailed information on the precautions to minimize the airport's environmental impact. For instance, he mentioned that they recycled items confiscated from passengers during security checks, such as water bottles and umbrellas. He highlighted that the airport was the only institution in the region treating wastewater before releasing it into the sea. Additionally, he discussed the types of fuels used by airplanes, the approximate quantity of fuel consumed during travel, and the methods employed to maintain surface friction on runways contaminated with snow and ice (noting that this has not been necessary thus far due to the area's limited snowfall during winters).

In response to preservice teachers' inquiries about the impact of airplanes on seagulls, the specialist explained that interference with aircraft takeoffs and landings is prevented by using sound to deter seagulls. The airport director further elucidated the mitigation process and outlined future operations, including unveiling the airport's upcoming international flight schedule.

When questioned about the criteria for selecting an appropriate location for airport construction, the director revealed several factors were considered. These considerations included a comprehensive analysis of meteorological data from the past 50 years, with particular attention to the peak days of the hottest and coldest temperatures. In addition, academic experts evaluated the potential harm to marine life resulting from the airport's construction and operation. The director also pointed out that the region is a popular destination for nature tourism. The establishment of an airport would, therefore, enhance accessibility for tourists.

During the final week of the study, participants were instructed to compose a concise essay delineating their ultimate decision regarding the viability of constructing the Rize-Artvin airport within the specified region. Additionally, respondents were required to complete the VNOS-B questionnaire as part of the evaluative process.

9.4.1 Data Analysis

Subsequent to the data collection phase, two different analytical approaches were employed for data analysis. Specifically, content analysis, a widely utilized method in qualitative research, was applied to scrutinize the concisely written essays wherein preservice teachers explicated the rationale underpinning their decisions concerning the Rize-Artvin airport. In the context of this study, content analysis was chosen as the analytical framework to investigate empirical documentation. Content analysis, as described by Berelson (1952), is characterized as a systematic and replicable technique designed to condense extensive textual content into fewer content categories, guided by explicit coding rules. Given that content analysis encompasses both qualitative and quantitative dimensions in the analysis of written, verbal, or visual communication messages (Elo & Kyngäs, 2008), it was deemed a fitting method for the study's objectives. The authors established and delineated criteria for the thematic analysis based on the responses provided by the students.

The answers to the VNOS-B survey were classified according to the established coding guidelines for the VNOS-B, as suggested by Lederman et al. (2002). Responses were labeled as 'naive' if they failed to correspond with the advised understanding of the NOS aspect. They were considered 'adequate' if they largely matched the NOS aspect and 'informed' if they not only aligned with but also provided detailed explanations and examples of the NOS aspect. For example, in response to the VNOS-B questions regarding the tentativeness, if preservice teachers responded, "Science never changes," their conception was coded 'naive.' If they responded, "Science changes based on new evidence," their response was coded as 'adequate.' If they responded, "Science changes with new evidence or through reevaluating existing evidence," their response was coded as 'informed' (Akerson et al., 2019).

9.5 Outcomes

Preservice teachers exhibited alterations in their decision-making methods before and after participating in the place-based SSI activity. This shift was evident in their essays, which focused on the decision to either support or reject the construction of the Rize-Artvin airport in the local region. Their essays provided insights into how their thought processes and decision-making strategies evolved through engagement with the SSI activity, particularly concerning environmental harm.

Prior to the Rize-Artvin airport excursion, three students possessed local knowledge of the construction owing to residency in the region, while others focused on topographical and meteorological characteristics of the region. Initial decisions from preservice teachers yielded 18 negative, four positive, and two undecided responses. In Table 9.2, we showed the frequencies of the codes and their frequencies in agreed and disagreed decisions and undecided cases. When constructing the table, specific codes were considered rather than individual counts, thus precluding an accurate reflection of the number of participants.

Table 9.2 reveals that individuals opposing the airport construction in the initial phase cited the region's unsuitability. This was based on factors such as mountainous terrain, unfavorable weather conditions, insecurity (vulnerability to natural disasters like earthquakes or flooding), and the region's low population density. In the responses provided before the activity, 27 preservice teachers expressed the opinion that an airport should not be constructed in the region due to its physical conditions. However, in the responses after the activity, only two students raised this concern.Two illustrative excerpts from the written reflections of preservice teachers are provided below.:

Table 9.2 Frequencies of the codes and their frequencies in agreed and disagreed decisions and undecided cases

Codes	Pre	Post
Disagree with airport construction		
Physical features of the region	27	2
Economic reasons	10	5
Impact on the environment	6	18
Agree with airport construction		
People's and Regio's welfare increased	1	15
Impact on the region's tourism and culture	2	5
Convenience in transportation	1	9
Undecided		
Another nearby airport exists, but commuting to this location poses challenges	1	–
Biodiversity suffered, but tourism benefited	1	1

The ideal location for constructing an airport necessitates a flat and expansive terrain. (Pre-PST1)

For optimal airport construction, the preference should be for flat terrain, as building on hilly landscapes incurs higher costs. A flat land ensures smoother operations, preventing obstacles for aircraft. Consequently, building an airport in Rize is not advisable due to its topography. (Pre-PST4)

Another reasoning articulated by the preservice teachers against the construction of an airport in the region pertains to the perceived economic impracticality, positing that the construction costs would outweigh the potential revenue generated by the airport. This sentiment was expressed by ten preservice teachers in their pre-activity essay responses and by five preservice teachers in their post-activity essay reflections.

Constructing airports in metropolitan areas is more sensible. This is because airport construction is expensive, and having fewer staff, such as pilots and hostesses, makes it even more costly for airports to operate with limited passengers. Metropolitan cities attract more passengers, making airport presence more logical. Consequently, given Rize's small size, it is not advisable for the city to have its own airport. (Pre-PST10)

I believe it was a bad idea. The environmental and social problems it caused influenced my opinion. Additionally, the lack of profitability in the airport investments reinforces this perspective. (Post-PST23)

Before the placed-based and argumentation activity, only six preservice teachers highlighted the airport's negative impact on nature and the environment. However, following the activities, a total of 18 PSTs expressed concerns about the airport's effects on the environment, nature, and the residing living creatures.

> Numerous factors are taken into account during the construction of the Rize-Artvin airport, with the wind factor being the most crucial. Windbreaks, in the form of shutters, were erected to mitigate the impact of the wind. (Pre-PST8)

> Building the airport harms the homes of nearby residents. To fill the sea and gather materials for construction, stones, and soil were extracted from the cultivated areas of nearby villages. This process is expensive and environmentally harmful. (Pre-PST12)

> I oppose building the airport because it harms the environment and seems unnecessary, leading to both environmental damage and inefficient use of funds. Additionally, the noise from the airport disrupts the surrounding homes. (Post-PST4)

Preceding the place-based and the subsequent argumentation activity, a greater proportion of preservice teachers (N = 18) held an unfavorable perspective regarding the construction of an airport within the region, with their corresponding reasoning outnumbering those in favor. However, following the completion of the visit and engagement in the activity, there was an observed elevation in the number of PSTs (N = 15) expressing support for the airport construction. Among the reasons articulated by proponents of the airport's development was the anticipation of enhanced regional welfare and economic prosperity resulting from the establishment of the airport. This shift in opinions underscores the potential influence of experiential learning and discourse-based activities in reshaping perspectives among pre-service teachers.

> I support the construction of the Rize-Artvin airport. The airport in Trabzon is insufficient, and building one in Rize is economically necessary. The existing airport in Trabzon is reaching its passenger limits, causing delays for travelers heading to Artvin who previously had to land in Batum, leading to significant time consumption. Additionally, the new airport contributes positively to employment, tourism, and trade in the region. (Post-PST10)

> The airport ought to be constructed as it contributes to the city's economy and boosts tourism. (Post-PST24)

Five PSTs expressed support for the construction of the airport, emphasizing its potential to boost tourism and cultural exchange in their post-essay responses. Two students had mentioned this reason in their pre-essay, and this perspective was notably influenced by information gleaned from airport authorities during the site visit:

> It is crucial to construct airports in all regions, as they bring significant benefits to local tourism and the economy. (Pre-PST11)

> I support building the airport because it has boosted the economy, tourism, and the overall region. While acknowledging potential drawbacks in any project, I believe the benefits outweigh the negatives. I anticipate its sustained success, especially given its accomplishment of serving 1 million passengers annually. (Post-PST17)

One positive reason was that the airport facilitates transportation. Initially, only one PST acknowledged this before the places-based visit and argumentation activity. However, after these activities, nine PSTs cited improved transportation as a reason.

Furthermore, through their research, the PSTs discovered that the airport plays a crucial role in swiftly transporting critically ill patients to major city hospitals, which they consider a positive aspect.

> I believe constructing the Rize-Artvin airport is essential. It is crucial for the city's economic growth and overall development, as it facilitates necessary transportation. Moreover, the airport ensures prompt access to larger hospitals for emergency medical situations, benefiting those in need, particularly individuals facing health challenges in the city. (Post-PST19)

Moreover, before the excursion, two PSTs were undecided about the airport's construction, and one PST remained undecided after the trip. They highlighted a balance between positive and negative aspects, considering the proximity of another airport, albeit with lengthy transportation, and the negative impact on species diversity, juxtaposed with the revitalization of the region.

The essay outcomes revealed that preservice teachers employed more evidence-based explanations in their decision-making essays following their research on the issue and place-based experience. The context of the sea-filled airport necessitated the use of evidence regarding the construction and operation of the airport, data interpretation, recognition of diverse perspectives, and the formulation of resolutions. This shift towards evidence-based reasoning was also reflected in their responses to the VNOS-B questionnaire, indicating a deeper understanding and application of critical thinking skills in real-world contexts.

9.5.1 *Preservice Teachers' NOS Conceptions Before and After the Implementation*

The study observed significant development in understanding NOS concepts among preservice teachers following a place-based implementation, as detailed in our study and summarized in Table 9.3. The participants demonstrated marked improvement in comprehending key NOS elements such as tentativeness, creativity, subjectivity, and distinguishing between opinions and scientific knowledge. Notably, the aspects of tentativeness, along with creativity and imagination, showed the most

Table 9.3 Pre-service teachers' pre and post-NOS conceptions

	Pre-implementation			Post-implementation		
NOS aspects	Naive	Adequate	Informed	Naive	Adequate	Informed
Tentativeness	9	13	0	0	7	15
Observation vs. inference	18	4	0	19	2	1
Empirical	16	5	1	13	3	6
Creativity and imagination	9	12	1	2	5	15
Subjectivity	9	11	2	4	4	14
Opinion vs. scientific knowledge	16	5	1	8	6	8
Theory and law	20	2	0	19	2	1

progress. For instance, 15 preservice teachers exhibited informed views on tentativeness in the post-test, a significant increase from none at the pre-test level. Similarly, only one had an informed view on creativity and imagination in the pre-test, which rose substantially in the post-test.

Subjectivity was another area where notable development was observed. By the end of the intervention, 14 preservice teachers had reached an informed view on this aspect, compared to only two at the pre-test. While the improvement in understanding the difference between opinions and scientific knowledge was less pronounced than the other aspects, it was still significant. Eight preservice teachers were classified as informed in the post-test, up from just one in the pre-test.

Despite these positive trends, there was no noticeable advancement in their comprehension of observations versus inferences or in differentiating between scientific theories and laws. This finding suggests that, although there was a substantial overall enhancement in the PSTs' grasp of several NOS aspects, the areas of observation versus inference and the distinction between theories and laws still demand more focused attention.

This finding aligns with expectations for several reasons. Firstly, the preservice teachers in the course were non-science majors, and the course was designed to exploration is how they plan and conduct activities based on learning objectives in an out-of-school environment. We implemented a place-based approach to illustrate how local SSI could be addressed through site visits, thereby improving students' comprehension of the issues, encouraging multiple perspective analysis, and promoting evidence-based evaluations. Secondly, our focus was not on conducting a range of NOS activities. We concentrated on discussing and reflecting upon those NOS aspects that were observable through the students' place-based visits and their research. Aspects like theory and law, which were less tangible in these visits, were not emphasized.

As previously elucidated, a subset of the preservice teachers in this study resides within the region and possesses firsthand experience with the airport filling process, either proximal or remote. Notably, one of these participants (PST12) articulated that the soil and stone fragments utilized for airport filling were sourced through excavation within their own villages, resulting in the denudation of local lands. Consequently, these individuals commenced the activities with a predisposition towards a negative perspective, overlooking the potential benefits of the airport. However, following the place-based and argumentation activities, this particular participant revised his stance, acknowledging the potential merits of the airport under the condition that it is not constructed along the coastline. On the other hand, pre-service teachers without regional residency predicated their initial decisions and supporting viewpoints upon the physical conditions of the locale. Subsequent to the activities, their perspectives shifted towards a more scientific and evidence-based foundation.

This approach is supported by previous research, which indicates that some NOS aspects are more challenging to grasp than others (Ağlarcı et al., 2016; Akerson et al., 2014; Avsar Erumit & Yuksel, 2023). For example, understanding theory and law within a short time frame is notably difficult, as highlighted in studies like Avsar

Erumit et al. (2019). Previous research also suggests that students are more adept at conceptualizing concrete NOS aspects, such as empirical data, while their grasp of more abstract concepts tends to develop later (Akerson et al., 2014). Therefore, our findings are consistent with these patterns in NOS education, particularly given the course constraints and the students' backgrounds. However, akin to inherent constraints present in any research endeavor, this study is not exempt from limitations. Future iterations of similar investigations may benefit from replication in diverse contexts, incorporating activities that address additional dimensions for a more comprehensive understanding.

9.6 Transfer to the Global Context

Sea-filled airports are typically constructed in areas with limited land availability, often in densely populated or geographically constrained regions. These airports can be found across the globe. Japan's Kansai International Airport, situated on an artificial island in Osaka Bay, is one example, built due to land scarcity in the densely populated cities of Osaka, Kobe, and Kyoto. Similarly, Hong Kong International Airport, also known as Chek Lap Kok Airport, is built on a large artificial island created from the flattened Chek Lap Kok and Lam Chau islands. Macau International Airport is another instance, with its runway on a strip of reclaimed sealand and the passenger terminal on the adjacent Taipa Island.

Place-based instruction focusing on airports can be implemented as case studies in regions where sea-filled airports exist, providing a valuable SSI context for the subject matter. When these airports do not exist in a particular region, they can still be used as an SSI case and examined in class using different pedagogical approaches. These approaches include but are not limited to argumentation (Kim et al., 2014; Martín-Gámez & Erduran, 2018; Rundgren et al., 2016), modeling (Avsar Erumit & Yuksel, 2023), and role-play (Avsar Erumit et al., 2023; Namdar & Oğuz Namdar, 2021), are commonly used in SSI instruction.

Moreover, the increasing number of airports in different parts of the world and their regional necessity can provide an SSI context and opportunities for place-based instruction, particularly when visiting the nearest airport. Additionally, land reclamation for various purposes, such as residential and commercial complexes – as exemplified by much of lower Manhattan in New York City – can be a practical example when discussing SSI.

9.7 Conclusion

Aviation operations are often viewed as controversial due to their contribution to global warming. Sea-filled airports, which require significant earthwork during construction and operations, present even more pronounced controversies, especially since they are built over marine ecosystems. This study examined the implementation of place-based learning in out-of-classroom learning environment classes for preservice teachers. The study encompassed various activities and group discussions where preservice teachers explored different aspects of the Rize-Artvin Airport, such as its economic, environmental, scientific, and societal implications. They collected evidence, formulated arguments, and presented their perspectives on the airport's regional necessity. Subsequently, the preservice teachers had the opportunity to visit the airport, interact with experts, and pose questions related to their research.

The study demonstrated how place-based learning can enhance preservice teachers' understanding of the complex relationship between science and society. By engaging in real-world experiences and examining SSI in a local context, the preservice teachers developed a more nuanced understanding of the nature of science. Further research is needed to examine the impact of place-based instruction that focuses on different SSI on students' understandings of contextualized NOS.

Analyzing pre- and post-essay responses revealed a significant evolution in the preservice teachers' understanding and opinions regarding the Rize-Artvin Airport. Our study revealed associations between the accuracy of students' views on the NOS after place-based SSI instruction. These findings align with those of Herman et al. (2019), who also found correlations between the accuracy and contextualization of students' post-place-based SSI instruction NOS views and the complexity of their explanations regarding trophic cascades.

Through active participation in place-based learning, preservice teachers gained profound insights into the complex decision-making processes involved in airport construction and operation. After implementing place-based learning, the preservice teachers developed more informed views on the collaborative efforts of multiple professional groups, such as meteorologists, scientists, environmental engineers, and geologists, in determining the location and construction of large-scale structures like airports.

The preservice teachers not only deepened their understanding of the subjective nature of scientific viewpoints but also recognized that various factors, including the economic interests of individuals and groups, can influence these viewpoints. They learned the importance of relying on empirical evidence to make well-informed decisions.

Overall, the place-based learning approach provided preservice teachers with invaluable insights into the sea-filled airport construction's intricate decision-making processes, fostering a comprehensive understanding of the interplay between scientific perspectives, economic considerations, and community interests.

References

Abd-El-Khalick, F., Summers, R., Brunner, J. L., Belarmino, J., & Myers, J. (2023). Development of *VAScoR*: A rubric to qualify and score responses to the views of nature of science (*VNOS*) questionnaire. *Journal of Research in Science Teaching, 1–48*. https://doi.org/10.1002/tea.21916

Ağlarcı, O., Sarıçayır, H., & Şahin, M. (2016). Nature of science instruction to Turkish prospective chemistry teachers: The effect of explicit-reflective approach. *Cogent Education, 3*(1), 1213350.

Akerson, V., Nargund-Joshi, V., Weiland, I., Pongsanon, K., & Avsar, B. (2014). What third-grade students of differing ability levels learn about nature of science after a year of instruction. *International Journal of Science Education, 36*(2), 244–276.

Akerson, V. L., Avsar Erumit, B., & Elcan Kaynak, N. (2019). Teaching nature of science through children's literature: An early childhood preservice teacher study. *International Journal of Science Education, 41*(18), 2765–2787.

Avsar Erumit, B., & Yuksel, T. (2023). Developing and using physical dynamic models on socioscientific issues to present nature of science ideas. *International Journal of Science and Mathematics Education, 21*(4), 1031–1056.

Avsar Erumit, B., Fouad, K. E., & Akerson, V. L. (2019). How do learner-directed scientific investigations influence students' questioning and their nature of science conceptions? *International Journal of Education in Mathematics, Science and Technology, 7*(1), 20–31.

Avsar Erumit, B., Namdar, B. & Oğuz Namdar, A. (2023). *Using place-based SSI instruction that utilize role-playing to promote preservice teachers' socioscientific accountability and NOS.* Paper presented at the 96th NARST International Conference, Chicago, IL, USA.

Bayik, C., & Abdikan, S. (2021). Monitoring of small-scale deformation at sea-filled Ordu-Giresun Airport, Turkey from multi-temporal SAR data. *Engineering Failure Analysis, 130*, 105738.

Bekdemir, Ü., & Özdemir, Ü. (2011). Doğu Karadeniz bölümünde gelişmekte olan yayla turizm merkezlerine bir örnek: Bektaş Yaylası/An example to developing centers of plateau tourism in Black Sea Region: Bektaş Plateau. *Doğu Coğrafya Dergisi, 7*(7), 9–35.

Bell, R. L., Matkins, J. J., & Gansneder, B. M. (2011). Impacts of contextual and explicit instruction on preservice elementary teachers' understandings of the nature of science. *Journal of Research in Science Teaching, 48*(4), 414–436.

Berelson, B. (1952). *Content analysis in communication research.* The Free Press.

Eastwood, J. L., Sadler, T. D., Zeidler, D. L., Lewis, A., Amiri, L., & Applebaum, S. (2012). Contextualizing nature of science instruction in socioscientific issues. *International Journal of Science Education, 34*(15), 2289–2315.

Elo, S., & Kyngäs, H. (2008). The qualitative content analysis process. *Journal of Advanced Nursing, 62*(1), 107–115.

Erdoğan, M., & Ercoşkun, O. (2021). Artvin Rize Havalimanı'nın ulaşıma ve çevreye etkileri. *Journal of Architectural Sciences and Applications, 6*(1), 250–267.

Herman, B. C. (2015). The influence of global warming science views and sociocultural factors on willingness to mitigate global warming. *Science Education, 99*(1), 1–38.

Herman, B. C. (2018). Students' environmental NOS views, compassion, intent, and action: Impact of place-based socioscientific issues instruction. *Journal of Research in Science Teaching, 55*(4), 600–638.

Herman, B. C., Owens, D. C., Oertli, R. T., Zangori, L. A., & Newton, M. H. (2019). Exploring the complexity of students' scientific explanations and associated nature of science views within a place-based socioscientific issue context. *Science & Education, 28*, 329–366.

Herman, B. C., Newton, M. H., & Zeidler, D. L. (2021). Impact of place-based socioscientific issues instruction on students' contextualization of socioscientific orientations. *Science Education, 105*(4), 585–627.

Herman, B. C., Poor, S. V., Oertli, R. T., & Schulte, K. (2023). Promoting young learners' NOS views through place-based SSI instruction. *Science & Education, 32*(4), 947–992.

Kadioglu, S. (2018). Research of the sea-filled airport OGU in the Black Sea, Turkey, using ground penetrating radar method. *Construction and Building Materials, 158*, 1123–1133.

Karisan, D., & Zeidler, D. L. (2017). Contextualization of nature of science within the socioscientific issues framework: A review of research. *International Journal of Education in Mathematics, Science and Technology, 5*(2), 139–152.

Khishfe, R. (2012). Nature of science and decision-making. *International Journal of Science Education, 34*(1), 67–100.

Khishfe, R. (2022). Nature of science and argumentation instruction in socioscientific and scientific contexts. *International Journal of Science Education, 44*(4), 647–673.

Khishfe, R., Alshaya, F. S., BouJaoude, S., Mansour, N., & Alrudiyan, K. I. (2017). Students' understandings of nature of science and their arguments in the context of four socio-scientific issues. *International Journal of Science Education, 39*(3), 299–334.

Kim, M., Anthony, R., & Blades, D. (2014). Decision making through dialogue: A case study of analyzing preservice teachers' argumentation on socioscientific issues. *Research in Science Education, 44*, 903–926.

Lederman, N. G. (2007). Nature of science: Past, present, and future. In S. K. Abell & N. G. Lederman (Eds.), *Handbook of research on science education* (pp. 831–879). Lawrence Erlbaum Associates.

Lederman, N. G., Abd-El-Khalick, F., Bell, R. L., & Schwartz, R. S. (2002). Views of nature of science questionnaire: Toward valid and meaningful assessment of learners' conceptions of nature of science. *Journal of Research in Science Teaching, 39*(6), 497–521.

Lee, D. S., Fahey, D. W., Skowron, A., Allen, M. R., Burkhardt, U., Chen, Q., et al. (2021). The contribution of global aviation to anthropogenic climate forcing for 2000 to 2018. *Atmospheric Environment, 244*, 117834.

Leung, J. S. C. (2022). A practice-based approach to learning nature of science through socioscientific issues. *Research in Science Education, 52*(1), 259–285.

Martín-Gámez, C., & Erduran, S. (2018). Understanding argumentation about socio-scientific issues on energy: A quantitative study with primary pre-service teachers in Spain. *Research in Science & Technological Education, 36*(4), 463–483.

Matkins, J. J., & Bell, R. L. (2007). Awakening the scientist inside: Global climate change and the nature of science in an elementary science methods course. *Journal of Science Teacher Education, 18*(2), 137–163.

Namdar, B., & Oğuz Namdar, A. (2021). Fostering students' values through role play about socioscientific issues. *The Physics Teacher, 59*(6), 497–499.

National Research Council. (2002). *For greener skies: Reducing environmental impacts of aviation*. National Academies Press.

Özür, N. (2018). Türkiye'de havalimanlarının kuruluş yerlerinin sürdürülebilir arazi kullanımı bakımından değerlendirilmesi. *Türk Coğrafya Dergisi, 71*, 15–25.

Rundgren, C. J., Eriksson, M., & Rundgren, S. N. C. (2016). Investigating the intertwinement of knowledge, value, and experience of upper secondary students' argumentation concerning socioscientific issues. *Science & Education, 25*, 1049–1071.

Sadler, T. D. (2004). Informal reasoning regarding socioscientific issues: A critical review of research. *Journal of Research in Science Teaching: The Official Journal of the National Association for Research in Science Teaching, 41*(5), 513–536.

Sadler, T. D., Chambers, F. W., & Zeidler, D. L. (2004). Student conceptualizations of the nature of science in response to a socioscientific issue. *International Journal of Science Education, 26*(4), 387–409.

Semken, S., Ward, E. G., Moosavi, S., & Chinn, P. W. (2017). Place-based education in geoscience: Theory, research, practice, and assessment. *Journal of Geoscience Education, 65*(4), 542–562.

Smith, G. A. (2002). Place-based education: Learning to be where we are. *Phi Delta Kappan, 83*(8), 584–594.

Transportation Research Board Report (2021). *Transportation research circular E-C271: Critical issues in aviation and the environment*. Retrieved 6 June 2023, from http://www.trb.org/Main/Blurbs/182198.aspx.

Yan, H. K., Wang, N., Yu, T. L., Fu, Q., & Liang, C. (2013a). Comparing effects of land reclamation techniques on water pollution and fishery loss for a large-scale offshore airport Island in Jinzhou Bay, Bohai Sea, China. *Marine Pollution Bulletin, 71*(1–2), 29–40.

Yan, H. K., Wang, N., Wei, L., & Fu, Q. (2013b). Comparing aircraft noise pollution and cost-risk effects of inland and offshore airports: The case of Dalian International Airport, Dalian, China. *Transportation Research Part D: Transport and Environment, 24*, 37–43.

Zeidler, D. L., & Nichols, B. H. (2009). Socioscientific issues: Theory and practice. *Journal of Elementary Science Education, 21*, 49–58.

Zeidler, D. L., Walker, K. A., Ackett, W. A., & Simmons, M. L. (2002). Tangled up in views: Beliefs in the nature of science and responses to socioscientific dilemmas. *Science Education, 86*(3), 343–367.

Zhao, B., Wang, N., Wu, L., & Yang, C. (2010, June). Environmental pollution of offshore airport and semi-Offshore airport. In *2010 4th International Conference on Bioinformatics and Biomedical Engineering* (pp. 1–4). IEEE.

Chapter 10
Exploring Preservice Science Teachers' Emotive Reasoning through Place-Based SSI Instruction: The Case of Gold Mining

Ümran Betül Cebesoy

10.1 Introduction

The ultimate goal of science education is to raise scientifically literate individuals who are able to make informed decisions to participate in public debates about scientifically-based issues (Dillon, 2009; Lederman et al., 2014; Leung et al., 2020). However, Zeidler and his colleagues (Zeidler & Lewis, 2003; Zeidler & Keefer, 2003; Zeidler et al., 2005) argue that moral reasoning, ethical consideration, and social and character development are also important characteristics of scientifically literate individuals which are sine qua non for functional scientific literacy. As a result, SSI have been an important venue for promoting scientific literacy for the last few decades (Leung et al., 2020; Zeidler et al., 2019). SSI framework aims to engage students in the decision-making process about ongoing social issues which have ethical and moral dimensions embedded in scientific contexts (Zeidler & Keefer, 2003; Zeidler et al., 2005). Many issues ranging from gene technologies to climate change fall within the context of SSI (Cebesoy, 2021a, b; Herman, 2018; Sadler & Zeidler, 2005). These issues influence our lives at personal, regional, or global levels (Lee et al., 2013; Zeidler, 2014). Some environmental issues, such as renewable energy, global warming, or genetically modified foods often include environmental considerations that individuals need to make informed decisions by considering different perspectives and nature itself (Herman, 2018; Herman et al., 2020). However, this can be possible when the appropriate context for nurturing students' experiences, beliefs, interests, and feelings for others who are impacted by these environmental issues is provided. Thus, the science education community

Ü. B. Cebesoy (✉)
Department of Mathematics and Science Education, Faculty of Education, Usak University, Usak, Türkiye
e-mail: umran.cebesoy@usak.edu.tr

B. Namdar, E. Karahan (eds.), *Socioscientific Issues Focused Teacher Education*, Contemporary Trends and Issues in Science Education 60,
https://doi.org/10.1007/978-3-031-55233-5_10

needs to provide educational experiences for these students (Herman et al., 2020). These kinds of educational experiences leverage students' pro-environmental actions, such as effectively engaging SSI, including environmental concerns (Herman et al., 2020, 2021). One point is that emotive considerations and a desire to engage with nature trigger committing pro-environmental actions (Herman et al., 2020). Here, teachers play a crucial role in SSI teaching where the students are engaged with SSI (Nielsen, 2020; Zeidler et al., 2005). This research aims to explore how future teachers develop and express different forms of emotive reasoning through a six-week place-based SSI instruction.

10.1.1 Emotive Reasoning in the Context of Place-based SSI

Affective factors other than personal, cognitive, and social factors influence how students respond to SSI. In this respect, emotions play a crucial role in forming students' responses (Sadler, 2004). The role of emotions has long been investigated in SSI research and reported to be an important factor that facilitates students' engagement and decision-making in SSI (Cebesoy & Chang Rundgren, 2023; Herman et al., 2020; Sadler & Zeidler, 2004, 2005). In earlier research, Sadler and Zeidler (2004, 2005) explored college students' ideas, reactions, and feelings regarding genetic engineering issues and reported that emotions and moral considerations are important factors in students' decision-making in SSI. These findings echo the results of latter studies in different countries (i.e., Dawson & Venville, 2009; Cebesoy, 2020; Cebesoy & Chang Rundgren, 2023). Exploring Australian upper secondary students' decisions in biotechnology issues, Dawson and Venville (2009) reported that the students frequently used emotive reasoning other than rational reasoning. Cebesoy (2020) also found similar findings with Turkish preservice science teachers. She reported that most preservice science teachers (nearly 79%) used moral reasoning (either consequence-based or principle-based) and moral emotion-based reasoning in genetics-related SSI.

Sadler and Zeidler (2004, 2005) suggested that emotive-based considerations need to be taken into account while designing educational interventions to enhance students' informal reasoning and decision-making. To tackle this suggestion, several intervention studies were found in the literature focusing on developing participants' emotive reasoning throughout SSI-based intervention. For instance, Lee et al. (2012) explored preservice teachers' character and value development by participating in an education program focusing on different SSI topics such as nuclear power generation, climate change, and embryonic stem cell research. Participants showed general care for the characters presented in the scenarios and tended to approach the scenarios with emotion and sympathy. Moreover, they tended to negotiate moral conflicts with emotive reasoning while ignoring scientific evidence and claims. One major finding revealed that participants' views generally tended to consider local and personal perspectives (prioritizing national or individual profits over the well-being of the entire world). In this sense, their compassion towards other nations and cultures was relatively alleviated. In another SSI-based intervention, Herman et al. (2020) explored the development of college students' emotive reasoning during a place-based environmental SSI. While participants showed little care (passive care) to none (apathy) for the well-being of people and nature before the intervention, their intense concerns (moderated concern) and moral compassion for the well-being of people and nature matured after the instruction. Exploring preservice teachers' reactions to abortion after participating in an SSI-based intervention, Cebesoy and Chang Rundgren (2023) revealed that preservice teachers both approving and disapproving abortion frequently used consequence-, principle-, and emotive-based moral considerations while making decisions in the abortion issue. Nearly half of the participants showed sympathy and/or empathy towards the characters in the scenarios. The decisions of teachers in both groups (approving and disapproving abortion) relied on emotions such as the suffering of parents and child, and feeling sorry for the characters. The abovementioned studies highlighted the significant role of emotions and emotive considerations in negotiating SSI. Besides, SSI-based interventions were seen as an avenue for achieving this. While SSI-based interventions were mainly used to develop participants' emotive reasoning in previous studies (e.g., Cebesoy & Chang Rundgren, 2023; Lee et al., 2012), another important venue for developing participants' emotive reasoning is the place-based SSI instruction.

Place-based SSI instruction focuses on promoting students' sense of place and attachment to the real "others" (both people and the environment) that are influenced by SSI and their resolution, resulting in the promotion of ecological and cultural sustainability (Herman et al., 2018). The premises of place-based SSI instruction vary: leveraging students' sense of place (Herman et al., 2020), developing socioscientific reasoning (Zeidler et al., 2019), developing a more sophisticated and contextualized nature of science understanding (Herman, 2018; Herman et al., 2019, 2023), expression of more intense and nuanced forms of emotive reasoning (Herman et al., 2020, 2021), and increased level of accountability to engage and resolve environmental SSI (Herman et al., 2020, 2021).

10.1.2 Emotive Reasoning in the Context of Preservice Teacher Education

Emotive reasoning patterns have been explored in preservice science teacher education since 2008 in Türkiye (e.g., Topçu, 2008). While earlier studies mainly focused on enlightening the role and types of informal reasoning (e.g., rationalistic, moral-emotive, intuitive) of preservice science teachers (Topcu et al., 2011) and the role of context (Topcu et al., 2010), the latter studies focused on both identifying the preservice teachers' informal (Cebesoy, 2021a; Namdar et al., 2020; Kalin & Namdar, 2022; Sağlam & Eroglu, 2022) and moral reasoning patterns (Cebesoy, 2020) and developing their informal reasoning through SSI-based instruction in various SSI contexts (Cebesoy, 2021b; Cebesoy & Chang Rundgren, 2023; Karisan & Cebesoy, 2021). The studies focusing on SSI-based intervention were either focused on the case of genetics-related SSI (Cebesoy & Chang Rundgren, 2023; Karisan & Cebesoy, 2021) or environmental SSI (Cebesoy, 2021b). In her study Cebesoy (2021b) implemented an SSI-based instruction for developing preservice primary teachers' moral reasoning by using local and nonlocal environmental issues about deforestation and sea pollution. The results revealed that the inclusion of environmental SSI in environmental education courses facilitated preservice primary teachers' ecocentric and anthropocentric moral reasoning skills. Another significant finding was that local SSI was more influential in preservice teachers' decisions in some cases. For instance, 64% of participants indicated that a local sea pollution case (an oil tanker accident happening in the Marmara Sea of Türkiye) affected their decision more when compared to a nonlocal sea pollution case (the Deepwater Horizon accident which happened in the USA). Participants showed more sympathy and empathy towards the people and the environment affected by local sea pollution cases. Cebesoy's (2021b) findings were similar to Lee et al.'s (2012) findings as participants tended to prioritize local and national well-being over the well-being of the entire world. Besides, Herman et al.'s (2020) study exploring the development of students' emotive reasoning after participating in a place-based SSI instruction reported that more students showed moderated concern (higher sense of emotive reasoning) and deep compassion for nature and people who experience difficulties arising from environmental SSI. These interesting findings motivated the researcher to use a local SSI to design a place-based SSI to get a deeper insight into preservice science teachers' emotive reasoning development. The ultimate aim of this chapter is to explore the development of preservice science teachers' emotive reasoning after participating in a place-based SSI instruction on a gold mining case.

10.2 SSI Context: Gold Mining in Usak City

Gold has been a center of attraction for many reasons, such as its use in jewelry, trade, investment, and power since ancient times. Due to its gold reserve potential, Türkiye is a gold-rich country (Ersoy, 2022; Kaya, 2009; Mineral Research and Exploration [MTA], 2016). Türkiye's gold production has increased yearly since 2001 from 1.4 tons to 45 tons (Ersoy, 2022). There are four steps in mining as (1) researching the valuable metal, (2) stripping (trees are cut in the mining area, and the fertile soil in which the living things live is stripped) and blasting (after stripping the fertile soil, blasting is done regularly up to the layer where the ore is located), (3) grinding (the ore containing rock material is grinded) and leaching (where cyanidation process is conducted) and (4) storage of waste (after absorbing the valuable metals, the soil with cyanide is washed to reduce the level of cyanide and then stored in the tailing ponds and tailing dams) (Turkish Foundation for Combating Soil Erosion [TEMA] Gold Mining with Cyanide Leach Report, nd).

The three most common techniques are used in gold mining flotation, heap leaching, and tank leaching (Ersoy, 2022). While flotation is considered a pre-enrichment process in gold mining (MTA, 2016), the leaching procedure is mainly used in processing gold (MTA, 2016; Thenepalli et al., 2019). The basic steps in this process involved (1) crushing the ore which contains valuable metals, (2) spreading the crushed ore over geomembrane-lined pads which are solid and water-resistant to prevent spreading the leaching solvent to the soil, (3) spraying leaching solvent like sulfuric acid or cyanide over the pads, and (4) collecting the valuable minerals dissolved into the solution. The valuable metals are then collected by using absorption methods (Manning & Kappes, 2016; MTA, 2016; Thenepalli et al., 2019). This process has both advantages and disadvantages due to its use of leaching solvents such as cyanide. It is a low cost technology that helps mining companies to harvest precious metals from the ore. It is also efficient in terms of being a quick method to harvest the metals which leads to proper growth in companies' capital (Manning & Kappes, 2016; Thenepalli et al., 2019). However, it also has disadvantages such as permeability, stability, leaks, recovery issues, and acid production (Manning & Kappes, 2016). Thus, each stage of gold mining is criticized. Even at the research stage which is the first stage of gold mining, thousands of trees are cut down, and agricultural areas are damaged (TEMA Gold Mining with Cyanide Leach Report, nd). However, the national reports (e.g., MTA, 2016) emphasize that gold mining is an important source of income due to its economic contribution to Türkiye. In line with these reports, governments have supported the operation of gold mines since the 1990s (Özen & Özen, 2018). Accordingly, the number of gold mines operated in Türkiye has increased since then (MTA, 2016). This also caused pro-mining movements such as protests such as Bergama (located in Izmir) and Esme (located in Usak city) protests against gold mines operated in the region (Özen & Özen, 2018).

I specifically chose the gold mining case for this study for several reasons: (1) My students (senior preservice science teachers) are enrolled in a science teacher education program in Usak where the Kisladagi Gold Mine is located, (2) the students are mainly from Usak or living nearby cities such as Manisa or Denizli, (3) In past and present, there have been protests and lawsuits against operating a gold mine in Kisladagi region, (4) the preservice teachers had geology course in their previous semester and had information about mines, (5) as the participants are senior preservice science teachers who will be graduated and will start their profession soon, they need to be aware of both SSI and teaching SSI in their classrooms, and finally (6) the Kisladagi gold mine case is a suitable SSI for place-based SSI instruction.

10.3 Implementation Procedure

In this section, the implementation procedure will be explained in detail. The implementation was conducted as part of an SSI-oriented elective course entitled 'Problems related to Science and Technology'. In this course, preservice science

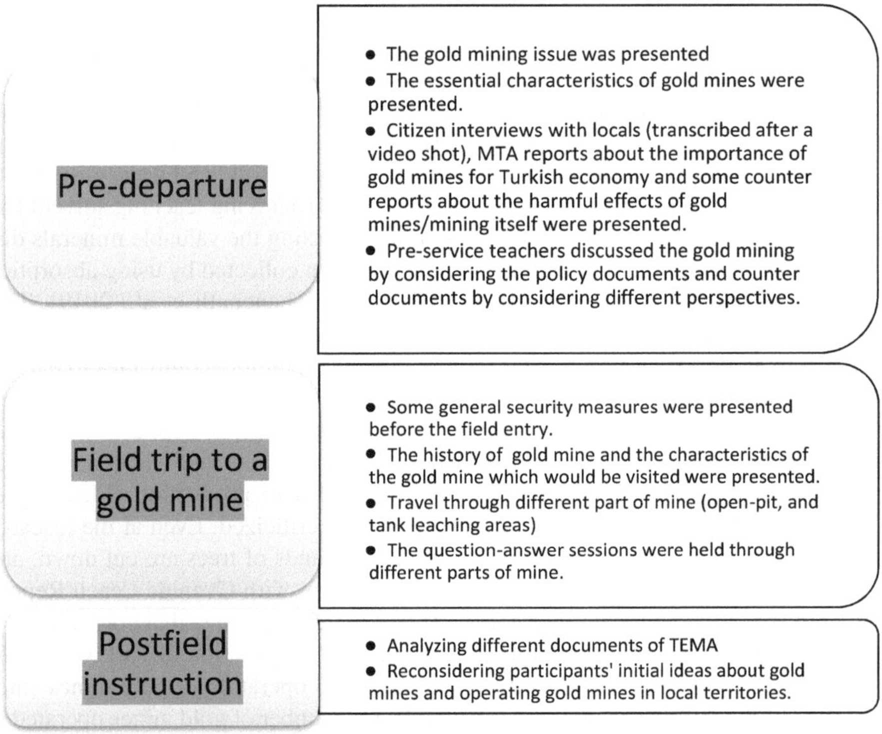

Fig. 10.1 The events held throughout the place-based SSI course

teachers got familiar with SSI, the characteristics of SSI, the role of multiple perspectives and scientific evidence in SSI, and the importance of reasoning during negotiating SSI. As part of this course, a 5-week place-based SSI focusing on gold mining was designed. This course was divided into three components: (1) predeparture instruction, (2) field experience, and (3) post-departure discussion (see Fig. 10.1).

10.3.1 Pre-Departure Instruction

The first phase of instruction took 2 weeks. It was important for preservice teachers to get familiar with the local mining issue and emotive reasoning before the field trip. Perspective-taking and emotive reasoning were introduced to teachers. Some of them did not even know whether there was a gold mine in Usak City. Some reports and articles about how gold is mined and how a gold mine is operated were presented throughout the 2 weeks. In addition, they explored one recent article entitled 'how gold mines destroyed sunflowers and olive growing fields' published in a newspaper (https://www.bbc.com/turkce/haberler-dunya-61257090). Some reports prepared by the Mineral Research and Exploration (MTA) about how gold is mined, different techniques used in gold mining, the importance of gold mining in Türkiye and the world, the reservoir of gold in Türkiye, and its contribution to the Turkish economy were presented. They were also exposed to locals' voices from an article exploring the peasant living near the Kisladagi gold mine (Özberk, 2022). In this article, interviews with locals who support and are against gold mining in Kisladagi were analyzed. This helped them to get familiar with different stakeholders' perspectives and the emotive facets surrounding the issue. Interactive discussions about why a gold mine should be or should not be operated throughout the territory where they are living and how gold should be mined were held.

10.3.2 Field Experience Instruction

After receiving the required permissions from the University, the Goal Mining Company, and the Department, a field trip to the gold mine was operationalized. This field trip is a one-day-long trip to an already operating gold mine. As the gold mine is a highly secured place, all the participants were given guest cards to prevent unauthorized entry. Here, first, the required security measures during the field trip were explained. All the participants were given protective glasses and vests, and hard heats before the trip. Also, an informative film about safety was presented by the senior geology engineer. The participants listened to how the gold mine was constructed back in 1993, and the first gold was mined in 1998. How the gold was extracted and how cyanide is used in gold mining were explained by the senior geology engineer. Besides, how work safety is provided for mine workers and how the buildings were built are explained. The company also received a 'Cyanide Code' which is an important certificate showing the safe management of cyanide by the company. After the theoretical part, there was a question-answer session in which the participants raised their concerns and asked their questions. The senior engineer discussed the concerns raised by the environmental organizations like handling the mine waste containing cyanide and the impact of cyanide on air and water quality. He explained the company's measures to prevent the waste containing cyanide from mixing with groundwater by presenting an actual geomembrane to the participants. Moreover, he showed some statistical reports showing the increase in the number of foxes, boars, and hedgehogs and how the Turkish government was nourished by this gold mining investment. In the last part, the company's investments in local territory like a building in the university, and dialysis center in the district, and various donations to the schools around the district were presented. Then, the participants along with the senior engineer visited the open-pit, and tank-leaching areas to understand how the gold is mined and processed after mining.

During the visit, the preservice teachers were immersed in nature, and the course instructor presented explicit prompts to the preservice teachers about nature.

10.3.3 Post-Field Instruction

The preservice teachers gained first-hand experience with how gold is mined and examined the gold mine and different parts of the gold mine. In addition, they had the opportunity to see the remains of a so-called mountain (Kışla Mountain). What they did actually see was a huge hole left from the mountain. After their visit, they revisited their decisions about whether they wanted a gold mine where they live. In the following week, the Provincial Representative of the Turkish Foundation for Combating Soil Erosion (TEMA) was invited to the course since it was the major non-govermental organization that stands against gold mining in Türkiye and had several ongoing legal cases and protests against gold mining companies in Türkiye. Due to some unforeseen reasons, the Provincial Representative could not attend the course but he sent some important policy documents such Mining in and around Afyon (Afyon ve Çevresinde Madencilik, TEMA, 2021), the Mining accident in İliç (a newspaper article indicating a mining company announcing a malfunction occurred in the pipelines carrying cyanide-containing solution in İliç, a province of Erzincan city), Areas closed to Mining Policy Note (Madenciliğe Kapalı Alanlar Politika Notu, TEMA, 2022) and Gold Mining with Cyanide Leach (Siyanür Liçi ile Altın Madenciliği, TEMA, n.d.). These documents along with videos are presented in Fig. 10.2. Four videos which were prepared by TEMA and uploaded to social media were used in this part: (1) "We have something to talk about gold mining!" (https://www.youtube.com/watch?v=5qdYFp_ZOXg), (2) "The people of Fatsa have something to talk about the gold mine!" (https://www.youtube.com/watch?v=KunkcZfSMvc), (3) "There is death below" (https://www.youtube.com/watch?v=ejjPv1NnHQk), and (4) "If the law does not protect, the mine does not let to live!" (https://www.youtube.com/watch?v=IlcHWyiFE1M).

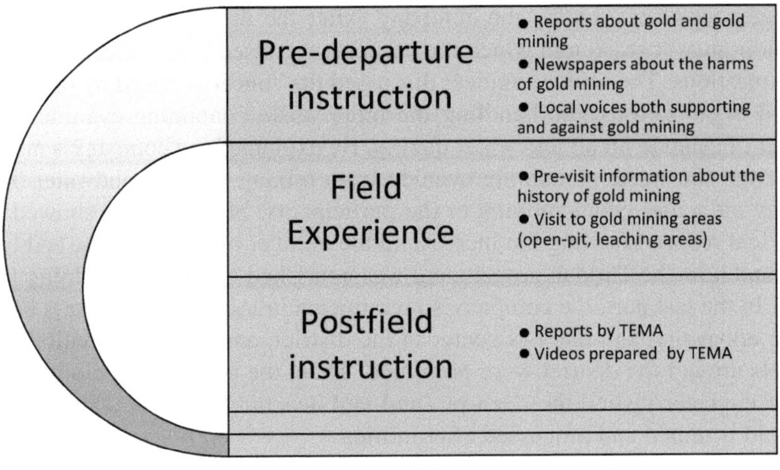

Fig. 10.2 The reports used during the place-based environmental SSI course

As the course followed, the preservice teachers had more chances to get familiar with different stakeholders' views (senior geology engineer's views, the workers' views, locals' views both supporting and against gold mining, TEMA representative's view and scientific documents on gold mining with cyanide). They discovered lots of interactions between nature and the people living in the territory. They also realized how gold mining with cyanide construes an environmental SSI, and different stakeholders' views were undeniable and important parts of the environmental SSI solution. They revisited their decisions during classroom discussions after new pieces of evidence, such as different documents, videos, or research findings, were introduced. In a final attempt about enlightening how their emotive reasoning changed during the course, the questionnaires along with interviews were conducted.

10.4 Outcomes

In this research, I explored how preservice teachers' emotive reasoning developed after participating in a place-based instruction about gold mining. For this purpose, I collected data from preservice teachers by using the SEEDS (Socioscientific and Environmental Engagement Survey). This questionnaire was administered before and after the course, and the questions were also used for interviews.

10.4.1 Socioscientific and Environmental Engagement Survey (SEEDS)

This survey was originally developed by Newton et al. (2016) by considering the items in Characters and Values as Global Citizens Assessment developed by Lee et al. (2013). The questionnaire includes 19 Likert- items and qualitative prompts in seven dimensions: Interconnectedness, sustainable development, moral and ethical sensitivity, perspective taking, empathic concerns, and examining scientific evidence. Each dimension has forced response Likert items accompanied by qualitative prompts which makes it easier to provide freely detailed comments that include numerous examples of empathy concerning various SEEDS dimensions (Herman et al., 2018).

SEEDS instrument is used in previous place-based SSI studies to explore participants' emotive reasoning (Herman et al., 2018), NOS views (Herman et al., 2019), and socioscientific operations (Herman et al., 2021). In this study, the SEEDS questionnaire was used to explore preservice teachers' emotive reasoning after participating in a place-based environmental education course. The participants completed the questionnaire as a pre-test before the instruction and a post-test after the course. A follow-up semi-structured interview with four participants lasting approximately 30 minutes was held after the course.

For this research, necessary permission to use the questionnaire was obtained from the author. Then, the forward translation procedure was applied (Hambleton, 2005). The questionnaire was translated into Turkish by the researcher, and two experts in SSI checked the language equivalency. As the researcher did not conduct any statistical analysis using the Likert-type items, and only used qualitative prompts (written responses to the SEEDS qualitative prompts) and interview data for data analysis, any further testing with Likert items (such as exploratory factor analysis or confirmatory factor analysis) was not conducted.

10.4.2 Data Analysis

To explore the development of preservice teachers' emotive reasoning throughout the course, all the qualitative data sources (the SEEDS questionnaire, interview data, and written responses to open-ended questionnaire items) were evaluated by using qualitative data analysis methods. The qualitative data were analyzed by using a semi-inductive approach (Miles & Huberman, 1994; Patton, 2015).

10.4.3 Participants

Fifteen preservice science teachers (four male and 11 female) from a public university located in the western Anatolia region of Türkiye participated in the study. The participants were senior preservice science teachers who were enrolled in an elective course entitled 'Problems related to Science and Technology' offered in the seventh semester of a science teacher education program in a medium-sized university located in the Western Anatolia region of Türkiye. All the PSTs were from middle-class socioeconomic backgrounds. Their age ranged from 21 to 23. Each participant signed an informed consent form before participation and was informed they were free to quit at any time during the study. An ID number (from 1 to 15) is given to each participant. For instance, PST-1 represents the first participant in the chapter.

10.4.4 Findings

Preservice teachers' expression of emotive reasoning and the development of their emotive reasoning after participating in a place-based SSI course were explored by using categories and subcategories of emotive reasoning. The categories and subcategories of emotive reasoning were developed by Herman et al. (2020). In their study, they revealed four major categories while answering the SEEDS questionnaire (1) apathy, (2) passive care, (3) moderated concern, and (4) empathetic dissonance. Under the *moderate concern* theme, participants showed a sentimental awareness and care for people and the environment that includes warnings and conditions. It has three subtypes (1) value judgment, (2) helplessness, and (3) diffusion of responsibility. When participants showed a deep emotional reaction identifying the people and nature that had negative effects, it is labeled as empathetic dissonance. It also has four categories: (1) Compassion, (2) guilt, (3) anger, and (4) righteous indignation. The major categories and subtypes of moderated concern and empathetic dissonance were presented in Table 10.1 along with operational definitions and exemplary responses that explain for each category best.

To explore how preservice teachers expressed emotive reasoning and how their emotive reasoning has changed after participating in a place-based SSI instruction, I used the categories and subcategories created by Herman et al. (2020). I, first, labeled preservice teachers' responses with the categories and subcategories of emotional reasoning. Then, I calculated the frequency of each type of reasoning before and after the place-based SSI instruction and compared them. Figure 10.3 presents preservice teachers' expressed types of moderated concern and empathetic dissonance. Note that I did not include apathy and passive care in this figure as both categories show less emotional attachment. Below, I presented how PSTs expressed different emotive reasoning patterns:

10.4.4.1 Apathy

As noted earlier, preservice teachers' statements lacking emotive attachment to the suffering of people and/or nature were coded under the apathy category. Some participants demonstrated an objective indifference towards SSI being discussed. Two excerpts exemplifying participants' indifference toward the SSI resolution are presented below:

> I did not see that most natural source problems were on the agenda. It happens rarely, but our agenda (referring to Türkiye's current situation) is mostly political and economic. So, I did not pay much attention to these issues. (PST-2, pre-course)

> I am undecided about whether scientific and technological developments will upset the balance in nature or not. Because technological progress is necessary for humans. I do not think that I have a personal attachment to nature at that point. (PST-11, pre-course)

Four of 15 preservice teachers showed apathy toward people and nature before the place-based SSI instruction. No one of the participants expressed apathy after completing the place-based SSI instruction.

10.4.4.2 Passive Care

Preservice teachers' statements showing a general concern for the welfare of people and nature were labeled as *passive care*. Participants' expressions in this category showed that they were cognitively sensitive to general principles and concerns

Table 10.1 Categories and subcategories of emotive reasoning along with excerpts

Categories of emotive reasoning	Subtypes	Operational definition	Sample excerpt
Apathy	–	Participants express emotional indifference to the suffering or mistreatment of people and nature.	I did not see that most natural source problems were on the agenda. It happens rarely, but our agenda (referring to Türkiye's current situation) is mostly political and economic. So, I did not pay much attention to these issues. (PST-2, pre-course)
Passive care	–	Participants show a general concern for the welfare of people and the nature	Interfering too much with nature does not produce good results for humanity. Although technological developments are good, they can upset the balance in nature. Nature takes back what it has given. (PST-4, pre-course)
Moderated concern	Value judgment	Participants express a value statement emphasizing the idea that those who were negatively impacted – people and nature – were somehow to blame because of their choices	Deep sinkholes were formed as a result of excessive use of groundwater in Konya (a city that is famous for sudden occurring sinkholes in Türkiye). This situation is irreversible for both agricultural areas and groundwater. I feel sorry for the destruction of natural resources in this regard, but I believe that the people there should bear the consequences because they irrigate uncontrollably. (PST-6, pre-course)
	Helplessness	Participants indicated they were unable to feel a deep concern for those who suffered either because of a lack of knowledge, insight, or influence	I often put myself in the shoes of both parties (referring to nature and the people) and often remain indecisive. On the one hand, I know that there will be great harm to nature. On the other hand, I think about the gain it will bring to people who work for a purpose. At the same time, since I know that the harm done to nature will harm human beings later on, I generally remain undecided on this issue. (PST-15, pre-course) I have only encountered natural source management issues on television and the Internet. I can understand them [referring to the people who had difficulties because of environmental SSI] and I feel sorry for them. But I cannot say I fully understand them since I have never experienced their burdens on my own. I cannot feel what they were going through. (PST-4, pre-course)
	Diffusion of responsibility	Participants express a statement diverting their responsibility away from themselves towards a larger community	Establishing factories and mines to contribute to the country's economy causes air, water, and soil pollution. The thought that we harm living things for the sake of money makes me sad. The state and the government should cover the financial burden of these people who live in agriculture and animal husbandry. (PST-3, pre-course)

Empathetic dissonance	Compassion	Participants showed a deep sense of empathy for those who suffer (people and nature) without considering what caused their pain	In the case of using and managing natural resources, I say if I were in their shoes. For example, if the gold mine was established in the district where I live and I had to migrate from my home because of this. It is a really sad situation. (PST-13, post-course)
	Guilt	Participants express feeling bad or ashamed for not taking action to implement sustainable solutions to lessen the harmful effects on people and nature.	I feel myself responsible for natural resource and environmental issues management. I personally pay attention to these issues. I warn the people around me. Although what I do affects 3–5 people, I think this number will increase even more. (PST-4, pre-course)
	Anger	Participants express their empathy for those who are suffering (environment and people) while openly expressing rage for a well-defined reason	I could have not imagined that mining gold was so damaging to the ecosystem. I think of if the gold mine belonged to me, would I still consider it normal? If I had not listened to different views on establishing and operating a gold mine, I would have thought differently about its establishment. But now I am against its establishment and operation. (PST-1, post-course)
	Righteous indignation	Participants expressed a strong emotional reaction to the suffering of people and nature as a result of alleged injustices and unfairness	Here, I put the different perspectives on a table and consider the pros and cons, and I prefer whichever is useful. In fact, I think that we should put ourselves in the place of not only living things with intelligence and will but also other living things. How would we feel if we saw ourselves as an ant, a rabbit, or a squirrel, and people came and harmed our living space? Let's adapt to the same situation to us. How would we feel if someone came to our house, forced us out, and destroyed the house? (PST-8, post-course)

Adapted from Herman et al. (2020)

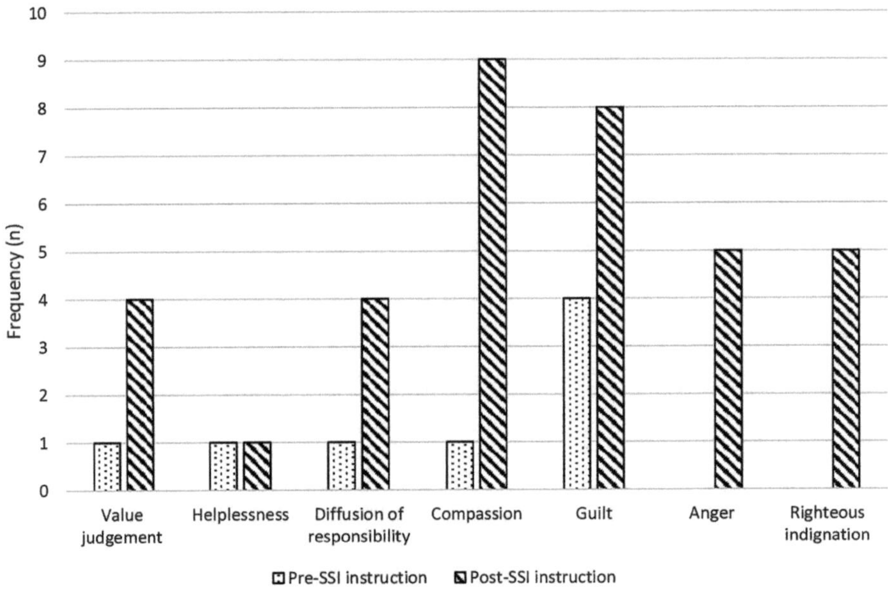

Fig. 10.3 Preservice teachers expressed forms of moderated concern and empathetic dissonance before and after the place-based SSI instruction

regarding environmental SSI by considering their initial values. Although passive care is a crucial step for developing strong emotive engagement, it lacks the emotional intensity that can provoke the person to take active steps to mitigate harmful effects. Sample statements in this category are presented below:

> Mines provide job opportunities and contribute to the country's economy. But the chemicals used (such as cyanide) are very harmful both to nature and to people living around them. (PST-6, pre-course)

> I think that the negative consequences of natural resource management can cause social and ethical conflicts. (PST-9, pre-course)

> Scientific and technological developments upset the balance of nature in some cases. Since people only think of their own future, the living spaces of other living things are restricted. This is causing the extinction of species around the world. (PST-10, pre-course)

While only four preservice teachers expressed general care about the welfare of people and nature before the instruction, all participants' expressions were coded as passive care showing participants had general concerns for the people and nature after SSI instruction to some degree. Only one participant continued to show a general concern for nature and the people after the instruction. He also supported establishing and managing the gold mine before, during, and after the SSI instruction. This is, in fact, an important detail since all preservice teachers who supported the idea of establishing and managing a gold mine where they live changed their minds after the instruction except this participant. He was consistent about his decision throughout the SSI-based instruction.

10.4.4.3 Moderated Concern

Moderated concern was noted when preservice teachers expressed emotional awareness and concern for the well-being of people and nature. However, they regulated their emotional concerns by adding conditions or limitations. The participants are less likely to engage in prosocial behavior when they regulate their emotional concerns. Participants in this study expressed three subtypes of moderated concern (1) value judgment, (2) helplessness, and (3) diffusion of responsibility. When Fig. 10.3 was examined, it could be seen that participants moderated their emotional responses toward people and nature three times before the place-based SSI instruction. The number of responses showing moderated concern was nine after the place-based SSI instruction indicating that there was an increase in participants' moderated concern-related responses.

10.4.4.3.1 Value Judgment

Value judgment is a form of moderated empathy and implies that the participants limited their emotional responses by indicating that those affected by environmental SSI were somehow to blame for their own suffering, as Herman et al. (2020) indicated. Two excerpts provided below illustrate how preservice teachers felt some concern toward people and natural resources and they regulated their concern by indicating that the people deserve their fate because of their choices and actions. PST-6 was the only one whose statement was coded under value judgment before place-based SSI instruction. She indicated that she felt sorry for people because of sudden sinkholes, but then regulated her emotive concern by indicating that the people deserved this fate because they irrigate the farmlands too much. Another participant (PST-4) indicated similar emotional consideration for the people living in the Black Sea region (a region that has recently been hit by sudden floods in Türkiye). She rationalized her concern by indicating that when people purposefully built their houses on the streambed in the region, it could be an expected result that sudden floods could harm those houses.

> Deep sinkholes were formed as a result of excessive use of groundwater in Konya (a city that is famous for sudden occurring sinkholes in Türkiye). This situation is irreversible for both agricultural areas and groundwater. I feel sorry for the destruction of natural resources and the people in this regard, but I believe that the people there should bear the consequences because they irrigate uncontrollably. (PST-6, pre-course)

> It makes me sad that people are harmed as a result of not managing natural resources correctly. For example, the houses built on the streambed in the Black Sea region were flooded. I felt sorry for them even though they did it consciously. (PST-4, post-course)

While only one (PST-6) moderated her concern by claiming that the people deserved it because of their acts before the place-based SSI instruction, the number of participants moderating their environmental concerns by indicating that the people impacted by environmental SSI deserved their fate increased after the place-based SSI instruction. Four PSTs moderated their concern toward people affected by environmental SSI as those people are responsible for their acts.

10.4.4.3.2 Helplessness

Helplessness is a way of moderating emotive reasoning. In this subtype of moderated concern, participants were unable to strongly empathize with the ones who were affected by environmental SSI because they were unable to relate to their suffering. The participants mentally and emotionally detached themselves from the people who are struggling with environmental SSI because of a lack of credible insight into the environmental SSI, lack of insight about the experiences of those people suffering, and ineffective attempts to resolve the issue (Herman et al., 2020). PST-4 was the only participant who indicated that even though she felt sorry for the suffering of people because of environmental SSI, she could not understand their exact feelings as she did not experience their difficulties.

> I have only encountered natural source management issues on television and the Internet. I can understand them [referring to the people who had difficulties because of environmental SSI] and I feel sorry for them. But I cannot say I fully understand them since I have never experienced their burdens on my own. I cannot feel what they were going through. (PST-4, pre-course)

We did not frequently encounter this type of moderated concern in the participants' responses either before or after the place-based SSI instruction. PST-6 was the only participant that indicated her emotive concern for the people and nature because of environmental SSI after place-based SSI instruction. She regulated her concern by claiming that she forgot it after a while as she did not physically experience the consequences herself.

> It makes me sad to hear and see some events from a distance. For instance, the sinkholes formed due to the excessive use of groundwater in agriculture in Konya [a city in Türkiye] caused a lot of damage to both the region and the people living there. This made me sad when I first heard about it. I thought about what would happen if it happened in the city where I live. But, then, I have forgotten as it does not cause any predictable harm to me. (PST-6, post-course)

10.4.4.3.3 Diffusion of Responsibility

One another way to moderate emotive considerations is the diffusion of responsibility. It refers to the moral responsibility to engage in prosocial behavior that can be affected when individual responsibility is veiled by group action, task allocation, and group decision-making (Herman et al., 2020). Two examples showing how participants moderated their concern by diffusing the responsibility of taking care of people who were affected by environmental SSI to third parties like the government or authorities are presented below. PST-3 indicated that she was feeling sorry for the people who live in agriculture and animal husbandry living around mines or factories. Then she diffused the responsibility of resolution of taking care of these people to the state and government by expressing that the state and government should cover their financial burdens. In a similar manner, PST-2 denoted that he felt sorry for the people living around the mine as this was not something those people desired. He relinquished the responsibility of taking care of these people to a third party as an authority.

> Establishing factories and mines to contribute to the country's economy causes air, water, and soil pollution. The thought that we harm living things for the sake of money makes me sad. The state and the government should cover the financial burden of these people who live in agriculture and animal husbandry. (PST-3, pre-course)

> This is actually a bad thing, no matter who it is [referring to the people who suffer from the results of natural resources management issues]. In the end, it is necessary to help them, as this is not a situation that occurs by their own will or mistake. Authorities need to take care of these people. If it were us, we would expect this kind of help. (PST-2, post-course)

When Fig. 10.3 was examined, it could be seen that one participant moderated their concern by diffusing the responsibility to resolve environmental SSI to third parties (e.g., authorities, government, state) before the place-based SSI instruction. After the course, four preservice teachers provided this type of moderated concern toward people affected by environmental SSI.

10.4.4.4 Types of Empathetic Dissonance

Empathetic dissonance is a strong emotional reaction to the welfare of the people and nature who are negatively affected by environmental SSI. This reaction is strong enough to potentially inspire participants to act in prosocial behavior to the resolution of environmental SSI (Herman et al., 2020). The preservice teachers showed four types of empathetic dissonance (1) compassion, (2) guilt, (3) anger, and (4) righteous indignation during place-based SSI instruction.

10.4.4.4.1 Compassion

Most participating preservice teachers demonstrated strong empathy for people affected by environmental SSI. However, they showed little interest in determining, blaming, or taking action to control the identified difficulties. Participants in this type of emotive reasoning identified some reasons for those difficulties but also indicated that they had no control over (such as nature) or that the difficulties raised were consequences of socially accepted norms such as laws and procedures. Below sample excerpts were presented. Participants in these excerpts expressed their compassion for the difficulties that the people who were living around the mine experienced such as losing their homes, and fertile fields.

> I both feel sorry for natural resource management and want a solution to it. I do not believe that anything will change by simply saying I feel sorry for the difficulties these people are experiencing. (PST-9, pre-course)

> I think we need to deal with those who are harmed by natural resource management. I am truly sorry for what they were going through. Because I think that agricultural work and cultural resources are more important than mining. People can live without wearing jewelry, but they cannot live without eating and drinking. (PST-12, post-course)

> In the case of using and managing natural resources, I say if I were in their shoes. For example, if the gold mine was established in the district where I live and I had to migrate from my home because of this. It is a really sad situation. (PST-13, post-course)

While only one participant (PST-9) showed compassion for the people affected by natural resource management in her response before the place-based SSI instruction, nine participants indicated similar concerns after the place-based SSI instruction. There is an increase in the number of participants who showed compassion for the people and nature being affected by environmental SSI after completing the course. Nine participants expressed some kind of compassion toward people and nature affected by environmental SSI.

10.4.4.4.2 Guilt

The statements coded under the guilt subtype of empathetic dissonance reflect participants feeling badly or ashamed if they did not participate in the resolution of environmental SSI ensuring sustainability. Sample quotations representing this type of empathetic dissonance are presented below:

> My conscience is not comfortable when I spend too much electricity and water. In this regard, I think that I am responsible to the society I live in, to the next generation, and even to the whole world. For example, if we waste water carelessly in this way, it will be a vital problem for the next generation. (PST-5, pre-course)

> I feel responsible for these issues [referring to social issues related to natural resource management and the environment] not directly but indirectly. I label myself as an environmentalist and a naturalist, but how much do I travel, see, examine, or research? Therefore, I feel indirectly responsible for everything that I am not aware of. I believe that this will be an important step if I raise the awareness of my students on this issue due to the profession which I will do [referring to being a teacher] and if I help them to get this awareness. (PST-8, post-course)

> I feel responsible for this [referring to social issues related to natural resource management and the environment]. I can do something against the pollution and deterioration of nature. I can go to environmental institutions and help them in their work. In this regard, I can invite my friends to these institutions. I can help them [referring to his friends] to be sensitive to the environment by constantly talking about these issues. (PST-14, post-course)

The quotations presented above exemplify how participants exhibit a high sense of moral agency and commitment to do prosocial behaviors such as using water and electricity less, increasing awareness of their students, and joining activities of environmental institutions. While there were four participants who showed a deep sense of guilt for not protecting the environment enough to leave it for the next generation before place-based SSI instruction, this number increased after the course. Eight out of 15 participants expressed this type of empathetic dissonance after completing the course.

10.4.4.4.3 Anger

This type of empathetic dissonance was noted when participants demonstrate a strong moral sense shown through obvious empathy for the suffering and explicit anger toward a specific cause of the pain (Herman et al., 2020). The sentiments presented below exemplify this type of empathetic dissonance. While PST-1 indicated that she was deeply affected by the effects of gold mines, PST-10 indicated her deep concern about excessive population growth on Earth and the damaging natural resources due to this growth.

> I could have not imagined that mining gold was so damaging to the ecosystem. I think if the gold mine belonged to me, would I still consider it normal? If I had not listened to different views on establishing and operating a gold mine, I would have thought differently about its establishment. But now I am against its establishment and operation. (PST-1, post-course)

> We are degrading the reputation of humanity by damaging natural resources. The primary reason is population growth. The fact that the population of the world is 8 billion now leads to some major problems such as food and shelter on this planet. The human population is inversely proportional to the other types of species. As the number of humans increases, other living species decrease. Suppose there are only humans on this planet. What a cruel life. I just could not imagine. (PST-10, post-course)

We did not observe anger in participants' responses before place-based SSI instruction. On the other hand, there were five participants who expressed their clear empathetic feelings toward nature and people who suffered from environmental SSI after the course.

10.4.4.4.4 Righteous Indignation

It is the last type of empathetic dissonance implying a strong emotional response to the suffering of people who were affected by environmental SSI. In addition, participants in this category referring to moral principles such as equity and justice were also violated (Herman et al., 2020). The sentiments below showed the indignation of participants while also implying strong emotional responses such as unjust treatment of other living things or mismanagement of natural resources.

> Here, I put the different perspectives on a table and consider the pros and cons, and I prefer whichever is useful. In fact, I think that we should put ourselves in the place of not only living things with intelligence and will but also other living things. How would we feel if we saw ourselves as an ant, a rabbit, or a squirrel and people came and harmed our living space? Let's adapt to the same situation to us. How would we feel if someone came to our house, forced us out, and destroyed the house? (PST-8, post-course)

> I honestly do not think people will feel comfortable conscientiously while considering social and ethical concerns raised in natural resources management. When natural resources are consumed and mismanaged, people do not feel comfortable conscientiously. As a human being, I am not comfortable. I am not alone in this world. We must leave the natural resources in the best accurate way to the next generations. (PST-1, post-course)

This kind of strong emotive consideration was not observed in participants' responses before place-based SSI instruction. Five participants expressed a sense of righteous indignation after place-based SSI instruction.

10.4.5 Summary of Chapter Outcomes

The findings revealed that preservice teachers mostly showed a general sense of care towards environmental SSI. Apathy indicating emotional indifference for the suffering or mistreatment of people and nature was seen before place-based SSI instruction (four out of 15 PSTs). This kind of emotional indifference was not observed after the instruction. Passive care reflecting general concern was observed in four participants' sentiments. All participants' expressions were coded as passive care showing PSTs had general concerns for the people and nature after SSI instruction to some degree. Participants' emotive reasoning reflecting moderated concern and empathetic dissonance varied between before and after place-based SSI instruction. There were only a few instances showing types of moderated concern before the place-based SSI instruction. The number of subtypes of moderated concern (e.g., value judgment and diffusion of responsibility) was slightly increased after the place-based SSI instruction. In a similar manner, there were few instances of the subtypes of empathetic dissonance (e.g., compassion and guilt) before the place-based SSI instruction. While only one participant showed a strong emotional response but neglected to determine, blame, or take action to control the identified difficulties (coded as compassion) before the instruction, nine participants showed this type of strong empathy after the course. Guilt was more frequently seen in participants' responses before the instruction. While there were four instances where participants showed signs of feeling badly or ashamed if they did not participate in the resolution of environmental SSI ensuring sustainability. These instances increased after the course. There were eight instances where I encountered participants' guilt-related responses. On the other hand, anger and righteous indignation were not seen in participants' responses before the place-based instruction. Five participants were showing obvious empathy for the suffering and explicit anger toward a specific cause of the pain. Similar to anger, five participants showed strong emotional responses to the suffering of people who were affected by environmental SSI while also indicating unjust treatment of other living things or mismanagement of natural resources.

As a teacher educator for more than a decade, I have been using cases for SSI instruction in my courses to develop preservice teachers' decision-making competencies, informal reasoning, and environmental moral reasoning (e.g., Cebesoy & Chang Rundgren, 2023; Cebesoy, 2020, 2021a, b). During this research, I myself realized that local issues intrigued participants' attention more (e.g., Cebesoy, 2021a, b), and started using local cases in my teaching. Planning and implementing a place-based SSI instruction by using a local gold mining case was both challenging and rewarding for me: First of all, planning and implementing a visit to a gold mine required many formal steps such as approvals from the department, faculty, and university of rectorate and similar steps were valid for gold mine. Since the mine is operated by a private company and is a highly secured place, the participants need to follow a visit schedule determined by the senior engineer and the company. They were quite helpful in getting official letters and helping the transferring of the participants from the university to the mine and backward by ensuring their own vehicles. No private car was allowed to access the mining zone. So, all the participants and I needed to follow their routine procedure. As a teacher educator and a participant, I saw the open-pit mining area with my students: a large deep hole (remaining of a so-called mountain-Kışla Mountain). During my implementation, I also used two different documentaries presenting interviews with locals living in nearby villages of the gold mine: one included interviews with locals supporting the gold mine, and the other included interviews with locals who were against the gold mine. I used these videos before the site visit. So, after participating in the site visit, the local voices became more meaningful. Provincial Representative (TEMA) was invited to the course since it was the major non-governmental organization that stands against gold and other valuable metals in Türkiye and had several ongoing legal cases and protests against mining companies in Türkiye at the post-field instruction. Unfortunately due to unforeseen reasons, he could not attend the course and this might hinder students' interest in the issue as they indicated that they were eagerly waiting for the representative's participation. Even though the planning and implementation process requires a lot of administrative work and unforeseen events, the participants' reactions to the SEEDS questionnaire and the mining issue were rewarding for me as many of them indicated they had never thought about how gold is extracted before and they had never assumed the local, economic, environmental and global dimensions in gold mining case before.

10.5 Transfer to the Global Context

Since mining has important potential for the economic growth of countries (MTA, 2016), it is a regular practice. However, the methods used in mining, such as using cyanide in leaching raise environmental concerns (TEMA Gold Mining with

Cyanide Leach Report, nd). Cyanide is a toxic material mainly used in the mining, galvanic and chemical industries. Cyanide ions are released into the environment mainly by wastewater (Jaszczak et al., 2017). As the soil with cyanide is washed and stored in the tailing ponds and dams, the storage areas create a major concern (Gold Mining with Cyanide Leach Report, nd). A recent review about cyanide in the environment reported that cyanide is dangerous for people's health and safety because the cyanide ions act as a poison. Moreover, the review also stressed its increased level of usage in the mining, galvanic and chemical industries (Jaszczak et al., 2017). Besides, using cyanide in extracting the gold from the ore, the mining industry itself has been heavily criticized by environmental organizations such as the Turkish Foundation for Combating Soil Erosion. Even the research stage is the first stage of gold mining where the gold repositories are searched, thousands of trees are cut down and agricultural areas are damaged (TEMA Gold Mining with Cyanide Leach Report, nd). However, Turkish governments have supported gold mining since 2000 (Özen & Özen, 2018). This is mainly due to its contribution to economic growth which is subject to many national policy documents published by Mineral Research and Exploration (MTA) (e.g., MTA, 2016). Due to its controversial nature, it has been an interesting and compelling issue for SSI researchers.

Promoting responsible citizenship and making science relevant for students are assumed to be overarching goals of the SSI framework (Evagorou & Dillon, 2020). Moreover, SSI is seen as a vehicle to facilitate the character and values of global citizens (Lee et al., 2012, 2013). The aforementioned studies reflect nurturing students' societal and global perspectives to make informed decisions as active citizens in a global society. Other studies used more local and personal issues while exploring students' decisions and reactions to SSI in an informed, fair and civic manner (Herman, 2015, 2018; Herman et al., 2019, 2020, 2021). For instance, the SSI program developed by Lee et al. (2013). Even though using different issues (global or local), the expectation was similar: SSI involvement at one scale (e.g., local and immediate) can be conceptualized on another scale (e.g., globally and extended) in contextually appropriate ways (Herman et al., 2021). For instance, engaging in a local SSI can promote students' character development and values as global citizens. Supporting this, the program aims to promote the character and values of students as global citizens, Lee et al. (2013) started by introducing the issues at the personal level to make them personally relevant and meaningful for students. In this respect, I used a local case for designing a place-based SSI instruction. Using the gold mining case as a compelling issue intrigued preservice teachers' attention in many ways: First, the preservice teachers who participated in the study were mainly from Usak and nearby cities where similar attempts for mining valuable metals such as gold, silver, and nickel are being processed (TEMA, 2021). Second, as gold itself has long been used for many purposes, such as jewelry or trade, the participants had not reflected upon its production stage until the course. Third, the gold company that participants visited is operated by a foreign international company which also captured participants' attention in terms of Türkiye's national benefits.

Using mines as a case has long been preferred in SSI research in different cultural contexts. In one of the earliest studies, Pedretti (1999) explored how Canadian sixth-grade students interpret and make a decision on a controversial mining topic by taking them to a science center exhibition about mining comprising four parts drilling the ore, blasting the surface, environmental effects of mining on wildlife and mining and daily life interaction. The students were asked to decide on building a mine in the town. The results revealed that taking students to an exhibition about mining enhanced their critical thinking and nurtured informed-decision making. In addition, students were more motivated and became emotionally attached since they presented arguments and rationales for their proposed actions on building a mine in the town. Moreover, using mines as a compelling issue stimulated dialogue and debate among students as well as evoked emotional attachment. Consequently, students reflected on different stakeholders, such as processes of science, the role of power, politics, and culture, and personal beliefs which resulted in informed decision-making at personal and societal levels (Pedretti, 2004). In a more recent study, Gutiérrez (2018) explored Colombian students' decisions on an oil extraction issue as a local case. The students showed empathy towards communities affected by mining projects. Suryawati et al. (2020) used gold mining as a local case in Indonesia and explored the effectiveness of a problem-based learning unit. They reported that Indonesian high school students' ability to identify, analyze, evaluate, and plan actions and sensitivity to local and global environmental issues was facilitated after the implementation. Gao et al. (2021) explored Korean high school students' emotional competence development after participating in an SSI program focusing on the Coltan mining problem and the designer baby dilemma. The implemented strategies were found to be effective in terms of nurturing students' empathetic abilities (e.g., empathetic listening), regulating their emotions, and enhancing students' emotional awareness. Moreover, the program affected students' empathy as well as also had a positive impact on students' character and values. SSI programs implemented in different countries using mining as a local case (Gao et al., 2021; Suryawati et al., 2020) have promising evidence for increasing students' emotional awareness and sensitivity and moral compassion toward environmental SSI. In addition, positive impacts on students' character and values as global citizens have been reported (Gao et al., 2021). These findings indicated that local issues could facilitate students' global citizenship.

While the aforementioned studies were conducted with students, similar efforts to enhance college students' emotive reasoning, and emotive competencies by using place-based SSI instruction were evident in literature (e.g., Herman, 2018; Herman et al., 2018, 2019, 2020, 2021). On the other hand, similar efforts in preservice science teacher education programs are relatively few (e.g., Lee et al., 2012). I tried to merge place-based SSI with a local mining case to enhance preservice teachers' emotive reasoning as teachers are key factors in implementing SSI in their classrooms (Tidemand & Nielsen, 2017; Nielsen, 2020). With this respect, place-based SSI instruction by using local issues as mining, in this case, could be transferred to different teacher education programs worldwide. Mining issues have gained attention from SSI researchers worldwide as a compelling issue for students (e.g., Gao et al., 2021; Gutiérrez, 2018; Pedretti, 2004; Suryawati et al., 2020). While the issue has the potential to evoke emotional attachment and to develop emotive reasoning, character, and values as global citizens, it can also be used for similar purposes in different teacher education programs worldwide. This chapter represents a case for achieving this. Place-based SSI instruction, as used in this chapter, can be implemented as part of elective courses. As the content of elective courses is determined by the teacher educators, place-based SSI instruction can be implemented throughout the course by considering careful preparation and some regulations for conducting the visits.

10.6 Conclusion

In this chapter, I explored the development of preservice science teachers' emotive reasoning after participating in a place-based SSI instruction by using local gold mining as a compelling issue. I presented instances where preservice science teachers developed their moderated concern by expressing emotional awareness and concern for the well-being of people and nature. In a similar manner, I presented examples where empathetic dissonance shows strong emotional reactions for the welfare of the people and nature that are negatively affected by environmental SSI. The number of preservice teachers who expressed moderated concern and empathetic dissonance has increased after participating in a place-cased SSI instruction. Using gold mine as a local and compelling issue attracted preservice teachers and provided opportunities for them to interact with nature and learn stories about actual people who were impacted by it (even preservice teachers watched their stories from documentaries presented to them) as well as learned about science (formation of gold on earth, methods used in mining, using cyanide for extracting the gold form the blast, storage of soil with cyanide, the methods used for storage and keeping it safe for a certain time) underlying environmental SSI. Such learning experiences are essential for cultivating preservice teachers' engagement and sociocultural and ecological sustainability (i.e., ecojustice), within local and global domains as Herman et al. (2020) indicated.

References

Cebesoy, Ü. B. (2020). Investigating Turkish pre-service science teachers' moral reasoning in genetics related socioscientific issues. *Journal of Science Learning, 4*(1), 20–30. https://doi.org/10.17509/jsl.v4i1.28155

Cebesoy, U. B. (2021a). Pre-service science teachers' informal reasoning patterns and risk perceptions in SSI: Case of gene therapy. *European Journal of Science and Mathematics Education, 9*(4), 211–229. https://doi.org/10.30935/scimath/11237

Cebesoy, U. B. (2021b). Investigating Turkish pre-service primary teachers' environmental moral reasoning patterns on environmental dilemmas. In A. Karmaoui, A. B. Salem, & M. H. Anees (Eds.), *Handbook of research on environmental education strategies for addressing climate change and sustainability* (pp. 39–62). IGI Global. https://doi.org/10.4018/978-1-7998-7512-3

Cebesoy, U. B., & Chang Rundgren, S. N. (2023). Embracing socioscientific issues-based teaching and decision-making in teacher professional development. *Educational Review, 75*(3), 507–534. https://doi.org/10.1080/00131911.2021.1931037

Dawson, V., & Venville, G. J. (2009). High-school students' informal reasoning and argumentation about biotechnology: An indicator of scientific literacy? *International Journal of Science Education, 31*(11), 1421–1445. https://doi.org/10.1080/09500690801992870

Dillon, J. (2009). On scientific literacy and curriculum reform. *International Journal of Environmental and Science Education, 4*, 201–213.

Ersoy, A. (2022). The current status of gold mining in Turkey: An overview. *Niğde Ömer Halisdemir University Journal of Engineering Sciences, 11*(4), 1103–1114.

Evagorou, M., & Dillon, J. (2020). Introduction: Socio-scientific issues as promoting responsible citizenship and the relevance of science. In M. Evagorou, J. A. Nielsen, & J. Dillon (Eds.), *Science teacher education for responsible citizenship: Towards a pedagogy for relevance through socioscientific issues* (pp. 1–11). Springer. https://doi.org/10.1007/978-3-030-40229-7

Gao, L., Mun, K., & Kim, S. W. (2021). Using socioscientific issues to enhance students' emotional competence. *Research in Science Education, 51*, 935–956. https://doi.org/10.1007/s11165-019-09873-1

Gutiérrez, M. F. (2018). Socioscientific argumentation and model-based reasoning: A study on mining exploitation in Colombia. *Universitas Psychologica, 17*(5), 1–12. https://doi.org/10.11144/Javeriana.upsy17-5.samb

Hambleton, R. K. (2005). Issues, designs, and technical guidelines for adapting tests into multiple languages and cultures. In R. K. Hambleton, P. F. Merenda, & C. D. Spielberger (Eds.), *Adapting psychological and educational tests for cross-cultural assessment*. Lawrence Erlbaum.

Herman, B. C. (2015). The influence of global warming science views and sociocultural factors on willingness to mitigate global warming. *Science Education, 99*(1), 1–38. https://doi.org/10.1002/sce.21136

Herman, B. C. (2018). Students' environmental NOS views, compassion, intent, and action: Impact of place-based socioscientific issues instruction. *Journal of Research in Science Teaching, 55*(4), 600–638. https://doi.org/10.1002/tea.21433

Herman, B. C., Sadler, T. D., Zeidler, D. L., & Newton, M. H. (2018). A socioscientific issues approach to environmental education. In G. Reis & J. Scott (Eds.), *International perspectives on the theory and practice of environmental education: A reader. Environmental discourses in science education* (Vol. 3, pp. 145–161). Springer. https://doi.org/10.1007/978-3-319-67732-3_11

Herman, B. C., Owens, D. C., Oertli, R. T., Zangori, L. A., & Newton, M. H. (2019). Exploring the complexity of students' scientific explanations and associated nature of science views within a place-based socioscientific issue context. *Science & Education, 28*, 329–366. https://doi.org/10.1007/s11191-019-00034-4

Herman, B. C., Zeidler, D. L., & Newton, M. (2020). Students' emotive reasoning through place-based environmental socioscientific issues. *Research in Science Education, 50*, 2081–2109. https://doi.org/10.1007/s11165-018-9764-1

Herman, B. C., Newton, M. H., & Zeidler, D. L. (2021). Impact of place-based socioscientific issues instruction on students' contextualization of socioscientific orientations. *Science Education, 105*(4), 585–627. https://doi.org/10.1002/sce.21618

Herman, B. C., Poor, S. V., Oertli, R. T., & Schulte, K. (2023). Promoting young learners' NOS views through place-based SSI instruction. *Science & Education, 32*(4), 947–992. https://doi.org/10.1007/s11191-022-00353-z

Jaszczak, E., Polkowska, Ż., Narkowicz, S., & Namieśnik, J. (2017). Cyanides in the environment—Analysis—Problems and challenges. *Environmental Science and Pollution Research, 24*, 15929–15948. https://doi.org/10.1007/s11356-017-9081-7

Kalin, B., & Namdar, B. (2022). Preservice science teachers' informal reasoning and scientific habits of mind: A case of hydroelectric power plants. *Turkish Journal of Education, 11*(1), 56–73. https://doi.org/10.19128/turje.980874

Karisan, D., & Cebesoy, U. B. (2021). Use of the SEE-SEP model in pre-service science teacher education: The case of genetics dilemmas. In W. A. Powell (Ed.), *Socioscientific issues-based instruction for scientific literacy development* (pp. 223–254). IGI Global. https://doi.org/10.4018/978-1-7998-4558-4.ch008

Kaya, M. (2009). Gold potential and production of Turkey. In *Proceedings-European Metallurgical Conference, EMC.*

Lederman, N. G., Antink, A., & Bartos, S. (2014). Nature of science, scientific inquiry, and socio-scientific issues arising from genetics: A pathway to developing a scientifically literate citizenry. *Science & Education, 23*, 285–302. https://doi.org/10.1007/s11191-012-9503-3

Lee, H., Chang, H., Choi, K., Kim, S. W., & Zeidler, D. L. (2012). Developing character and values for global citizens: Analysis of pre-service science teachers' moral reasoning on socioscientific issues. *International Journal of Science Education, 34*(6), 925–953. https://doi.org/10.1080/09500693.2011.625505

Lee, H., Yoo, J., Choi, K., Kim, S. W., Krajcik, J., Herman, B. C., & Zeidler, D. L. (2013). Socioscientific issues as a vehicle for promoting character and values for global citizens. *International Journal of Science Education, 35*(12), 2079–2113. https://doi.org/10.1080/09500693.2012.749546

Leung, J. S., Wong, K. L., & Chan, K. K. (2020). Pre-service secondary science teachers' beliefs about teaching socio-scientific issues. In M. Evagorou, J. A. Nielsen, & J. Dillon (Eds.), *Science teacher education for responsible citizenship: Towards a pedagogy for relevance through socioscientific issues* (pp. 21–39). Springer. https://doi.org/10.1007/978-3-030-40229-7

Manning, T. J., & Kappes, D. W. (2016). Heap leaching of gold and silver ores. In *Gold ore processings* (pp. 413–428). Elsevier. https://doi.org/10.1016/B978-0-444-63658-4.00025-6

Miles, M. B., & Huberman, A. M. (1994). *Qualitative data analysis: An expanded sourcebook* (2nd ed.). Sage Publications.

Mineral Research and Exploration (MTA). (2016). *Türkiye ve dünyada altın [Gold in Türkiye and in the world].* Retrieved from https://www.mta.gov.tr/v3.0/sayfalar/bilgi-merkezi/maden-serisi/img/Altin.pdf

Namdar, B., Aydin, B., & Raven, S. (2020). Preservice science teachers' informal reasoning about hydroelectric power issue: The effect of attitudes towards socio-scientific issues and media literacy. *International Journal of Research in Education and Science (IJRES), 6*(4), 551–567.

Newton, M. H., Herman, B. C., & Zeidler, D. L. (2016, January 5). *The longitudinal association of an experiential environmental SSI course with student conceptualizations and behaviors.* [Conference presentation]. Association for Science Teacher Education Conference, Reno, NV, United States.

Nielsen, J. A. (2020). Teachers and socioscientific issues – An overview of recent empirical research. In M. Evagorou, J. A. Nielsen, & J. Dillon (Eds.), *Science teacher education for responsible citizenship: Towards a pedagogy for relevance through socioscientific issues* (pp. 13–20). Springer. https://doi.org/10.1007/978-3-030-40229-7

Özberk, N. (2022). Anaakım ve alternatif medyada 'çevresel değerleme dili': Kışladağı altın madeni örneği ['Environmental valuation languages' in the mainstream and alternative media: The case of kışladağı gold mine]. *Kent ve Çevre Araştırmaları Dergisi, 4*(1), 36–59. https://doi.org/10.48118/yykentcevre.1062319

Özen, H., & Özen, Ş. (2018). What comes after repression? The hegemonic contestation in the gold-mining field in Turkey. *Geoforum, 88*, 1–9. https://doi.org/10.1016/j.geoforum.2017.11.002

Patton, M. Q. (2015). *Qualitative evaluation & research methods* (4th ed.). Sage Publications.

Pedretti, E. (1999). Decision making and STS education: Exploring scientific knowledge and social responsibility in schools and science centers through an issues-based approach. *School Science and Mathematics, 99*(4), 174–181. https://doi.org/10.1111/j.1949-8594.1999.tb17471.x

Pedretti, E. G. (2004). Perspectives on learning through research on critical issues-based science center exhibitions. *Science Education, 88*(S1), S34–S47. https://doi.org/10.1002/sce.20019

Sadler, T. D. (2004). Moral and ethical dimensions of socioscientific decision-making as integral components of scientific literacy. *The Science Educator, 13*, 39–48.

Sadler, T. D., & Zeidler, D. L. (2004). The morality of socioscientific issues: Construal and resolution of genetic engineering dilemmas. *Science Education, 88*, 4–27. https://doi.org/10.1002/sce.10101

Sadler, T. D., & Zeidler, D. L. (2005). Patterns of informal reasoning in the context of socioscientific decision- making. *Journal of Research in Science Teaching, 42*(1), 112–138.

Sağlam, H. I., & Eroglu, B. (2022). A mixed-method study on pre-service teachers' informal reasoning regarding nuclear energy use. *Journal of Turkish Science Education, 19*(2). https://doi.org/10.36681/tused.2022.139

Suryawati, E., Suzanti, F., Zulfarina, Z., Putriana, A. R., & Febrianti, L. (2020). The implementation of local environmental problem-based learning student worksheets to strengthen environmental literacy. *Journal Pendidikan IPA Indonesia, 9*(2), 169–178. https://doi.org/10.15294/jpii.v9i2.22892

Thenepalli, T., Chilakala, R., Habte, L., Tuan, L. Q., & Kim, C. S. (2019). A brief note on the heap leaching technologies for the recovery of valuable metals. *Sustainability, 11*(12), 3347. https://doi.org/10.3390/su11123347

Tidemand, S., & Nielsen, J. A. (2017). The role of socioscientific issues in biology teaching: From the perspective of teachers. *International Journal of Science Education, 39*(1), 44–61. https://doi.org/10.1080/09500693.2016.1264644

Topçu, M. S. (2008). *Preservice science teachers' informal reasoning regarding socioscientific issues and the factors influencing their informal reasoning* [Unpublished doctoral dissertation]. Middle East Technical University.

Topcu, M. S., Sadler, T. D., & Yilmaz-Tuzun, O. (2010). Preservice science teachers' informal reasoning about socioscientific issues: The influence of issue context. *International Journal of Science Education, 32*(18), 2475–2495. https://doi.org/10.1080/09500690903524779

Topcu, M. S., Yılmaz-Tüzün, Ö., & Sadler, T. D. (2011). Turkish preservice science teachers' informal reasoning regarding socioscientific issues and the factors influencing their informal reasoning. *Journal of Science Teacher Education, 22*(4), 313–332. https://doi.org/10.1007/s10972-010-9221-0

Turkish Foundation for Combating Soil Erosion (TEMA). (2021). *Afyon ve çevresinde madencilik [Mining in and around Afyon]*. Retrieved from https://cdn-tema.mncdn.com/Uploads/Cms/afyon-ve-cevresinde-madencilik_1.pdf

Turkish Foundation for Combating Soil Erosion (TEMA). (2022). *Madenciliğe Kapalı Alanlar Politika Notu [Areas closed to Mining Policy Note]*. Retrieved from https://cdn-tema.mncdn.com/Uploads/Cms/madencilige-kapali-alanlar-politika-notu_2.pdf

Turkish Foundation for Combating Soil Erosion (TEMA). (n.d.). *Siyanür liçi ile altın madenciliği [Gold Mining with Cyanide Leach]*. https://cdn-tema.mncdn.com/Uploads/Cms/siyanur-lici-altin-madenciligi.pdf

Zeidler, D. L. (2014). Socioscientific issues as a curriculum emphasis: Theory, research and practice. In N. G. Lederman & S. K. Abell (Eds.), *Handbook of research on science education, volume II* (pp. 697–726). Routledge.

Zeidler, D. L., & Keefer, M. (2003). The role of moral reasoning and the status of socioscientific issues in science education: Philosophical, psychological and pedagogical considerations. In D. L. Zeidler (Ed.), *The role of moral reasoning on socioscientific issues and discourse in science education* (pp. 8–38). Springer. https://doi.org/10.1007/1-4020-4996-X_15

Zeidler, D. L., & Lewis, J. (2003). Unifying themes in moral reasoning on socioscientific issues and discourse. In D. L. Zeidler (Ed.), *The role of moral reasoning on socioscientific issues and discourse in science education* (pp. 289–306). Springer. https://doi.org/10.1007/1-4020-4996-X_15

Zeidler, D. L., Sadler, T. D., Simmons, M. L., & Howes, E. V. (2005). Beyond STS: A research-based framework for socioscientific issues education. *Science Education, 89*(3), 357–377. https://doi.org/10.1002/sce.20048

Zeidler, D. L., Herman, B. C., & Sadler, T. D. (2019). New directions in socioscientific issues research. *Disciplinary and Interdisciplinary Science Education Research, 1*(1), 1–9. https://doi.org/10.1186/s43031-019-0008-7

Chapter 11
Codesign of Technology-Enriched Place-Based SSI Lesson Plans

Ferah Özer and Çiğdem Han Tosunoğlu

11.1 Introduction

SSI are complex and often controversial issues that involve both scientific and societal aspects. They require scientific knowledge and critical thinking skills to be addressed effectively. Technology integration has emerged as a promising approach to enhance students' engagement and learning outcomes in SSI in recent years. The following section discusses the importance of technology integration within the scope of preservice teacher education regarding SSI teaching.

11.1.1 Technology Integration in Teaching SSI

Technology integration in teaching SSI has been shown to offer several benefits. One of the key benefits is that it enhances students' engagement and motivation in SSI. Technology allows students to take an active role in the learning process and provides them with opportunities to explore and express their ideas efficiently. This, in return, leads to a deeper understanding of SSI, the related content knowledge, and a greater sense of ownership over the learning process. Additionally, technology integration promotes the development of collaboration and communication skills,

F. Özer (✉)
Office of Learning & Teaching (KOLT), Koç University, Istanbul, Türkiye
e-mail: feozer@ku.edu.tr

Ç. Han Tosunoğlu
Department of Mathematics and Science Education, Faculty of Education, Marmara University, Istanbul, Türkiye
e-mail: cigdem.han@marmara.edu.tr

B. Namdar, E. Karahan (eds.), *Socioscientific Issues Focused Teacher Education*, Contemporary Trends and Issues in Science Education 60,
https://doi.org/10.1007/978-3-031-55233-5_11

essential for addressing complex issues requiring multiple perspectives (Zorfass & Rivero, 2005).

Technology integration also provides access to a wide range of resources supporting teaching and learning in SSI (Chang et al., 2013). For instance, online simulations and models can assist students in understanding complex scientific concepts, whereas social media platforms and discussion forums can facilitate communication and collaboration between students from different locations and backgrounds. Technology can also provide opportunities for students to engage in authentic research and data analysis practices, helping them develop critical thinking and problem-solving skills (Yates et al., 2021). It is stated in the literature that there are several ways and tools for integrating technology into teaching SSI in the classrooms, and these are stated as follows:

1. Use of Online Resources: A variety of online resources are available that can help students explore SSI in more depth. These may include interactive simulations, online databases, and multimedia resources.
2. Use of Social Media Platforms: Social media platforms can be powerful tools for engaging students in SSI. These platforms, such as Twitter and Facebook, can be used to create classroom discussion forums and threads or to share news articles or research studies related to the issue.
3. Use of Videoconferencing: Videoconferencing tools such as Zoom, Microsoft Teams, or Google Meet can be used to connect students with each other and experts in the field, such as scientists or policymakers, who can provide insights and answer students' questions.
4. Use of Data Visualization Tools: Data visualization tools, such as infographics or interactive maps, can be used to help students understand and organize complex data related to SSI.
5. Use of Online Collaboration Tools: Online collaboration tools, such as Google Docs or Padlet, can be used to facilitate group work and discussion, even if students are not physically present in the classroom.

Overall, integrating technology and digital tools can help make teaching SSI more engaging, interactive, and relevant to students' lives and can provide them with the skills and knowledge they need to become informed and responsible citizens (Chang et al., 2013). It also diversifies teachers' teaching practices and enriches their repertoire while also increasing students' attitudes toward content knowledge (Christensen, 2002). On the other hand, while technology integration in teaching SSI has many benefits, it also has several challenges and barriers obstructing teachers from efficiently integrating them. One of the primary challenges is teachers' lack of experience in technology integration (Alkhawaldeh & Menchaca, 2014; Mishra & Koehler, 2006). In this context, it becomes even more important for preservice teachers to have experience with technology integration and digital tools during their school years as part of their teacher training programs to be fully equipped and efficient when they become in-service teachers.

11.1.2 Importance of Preservice Teacher Education

Research showed that the development of preservice and in-service teachers' understanding of SSI and their classroom practices are supported by various courses and activities such as science teaching methods courses, professional development programs, workshops, training, and argumentation-based activities (Borgerding & Dagistan, 2018; Cebesoy & Chang Rundgren, 2023). Although studies have shown that preservice teachers' pedagogical knowledge, argumentation skills, technology integration, motivation, and content knowledge about SSI have improved as a result of these initiatives, it remains unclear how preservice teachers integrate SSI into their classroom practices (Carson & Dawson, 2016; Saunders & Rennie, 2013). Moreover, preservice teachers state that they have difficulty finding a place to apply new approaches in schools within the scope of their practicum courses and that they need to be provided with collaborative opportunities with their peers while planning their courses (Topçu & Çiftçi, 2023). It is also evident from the studies that more effective results are achieved when preservice teachers collaboratively and actively participate in the process of planning and material development along with theoretical training in terms of integrating SSI practices into their teaching. (Tondeur et al., 2012).

At this point, collaborative teaching material design (co-design) approaches could be used in a supportive way during preservice teacher education. Co-design methodologies refer to collaborative, team-oriented processes that involve teachers, researchers, and other contributors working together closely to create innovative prototypes of teaching materials and assess their suitability for tackling educational challenges (Kwon et al., 2014). Collaboration is known as a process in which individuals work together to achieve a common goal (Webb, 2009). In the context of teaching material design and development, by working together collaboratively, the diverse skills and knowledge of each design team member are given equal importance, fostering mutual learning through their interactions. Thus, co-design for teaching material development has been recognized as an effective approach to enhancing the quality of teaching and learning in education (Kwon et al., 2014; Friedrichsen et al., 2020). Voogt et al. (2015) found that co-design enables instructors to design more effective teaching strategies, improve the quality of instructional materials, and better meet the needs of diverse learners. Co-design also helps teachers be more reflective about their teaching practices and develop a sense of ownership over the curriculum. Co-design can also improve communication and collaboration among teachers, leading to the development of a shared pedagogical vision (Voogt et al., 2016).

According to the studies, co-design processes can take many different forms. For instance, teachers can work together to design curriculum materials, adapt existing materials to meet the needs of their students, or design completely new materials (Slavin, 2014). It can also be undertaken in a lesson study format where teachers of the same discipline gather to design a mutual lesson plan, implement one after another under the surveillance of a fellow teacher, and provide constant constructive feedback in an iterative way (Özer et al., 2018). Although it requires additional effort, careful planning, and management to ensure the process is effectively handled over time, co-design can be a powerful tool for improving student learning and teacher motivation (Hancock et al., 2019). Teacher education programs provide a great learning space for students to collaboratively develop teaching materials step-by-step and improve the quality of their materials through constant and collaborative feedback from their peers and faculty members working with them. This process will help preservice teachers be equipped with the necessary teaching material design skills that they will need throughout their professional careers.

In addition, in recent years, it has been determined that SSI studies, especially with preservice teachers, generally examine the relationships between SSI and different variables such as pedagogical content knowledge, epistemological beliefs, informal reasoning, and attitudes (Cebesoy & Öztekin, 2016; Han-Tosunoglu & Özer, 2022; Ozturk & Yilmaz-Tuzun, 2017). Similarly, studies conducted with teachers have also focused on the determination of their understanding of SSI (Han-Tosunoğlu & İrez, 2017). However, studies on preservice teachers' and teachers' development of SSI teaching materials and/or implementation of SSI in the classroom are relatively limited. These few studies reveal that both preservice (Yapıcıoğlu, 2016; Genel & Topçu, 2016) and in-service teachers (Kutluca, 2021) have challenges in teaching SSI due to a lack of pedagogical content knowledge to teach SSI effectively; and lack of content knowledge about how to encourage students to discuss these issues and lack of experience in development of SSI lesson plans. In line with that, as Genel and Topçu (2016) stated, either preservice teachers do not put SSI practices on the agenda as an important topic, or the topics were discussed in a short time, and they did not address important dimensions of SSIs (for example, the moral dimension). Thus, studies indicate that preservice teachers should be supported in teaching SSI, and this support should not be limited to only providing theoretical background and aspects of SSI but also provide opportunities for them to experience designing SSI learning environments, lesson plans and discussing these issues in the real classroom environments, such as inclusion of co-design methodologies. However, from the context of SSI teaching, the inclusion of co-design approaches is yet an uncharted area, only a few studies included co-design implementation in the SSI context (Friedrichsen et al., 2020). Thus, more studies are needed to determine the effectiveness of these collaborative approaches.

11.1.3 The Purpose of the Study

As noted above, one of the major challenges for teachers is the insufficient familiarity and lack of experience with incorporating technology into SSI. It's important to consider that in-service and preservice teachers may not have the expertise and resources to make the most of these digital tools and technology. Many researchers (Holmes, 2009; Mishra & Koehler, 2006) pointed out that in-service and preservice teachers' teaching skills and repertoire should enable them to integrate technology in meaningful ways to improve student engagement, learning, and the quality of their teaching practices. Thus, they may require assistance and additional support to effectively utilize these technologies and tools in their teaching practices, specifically with SSI. Considering these, this study aims to provide support in the form of a professional development (PD) program to preservice biology teachers and examine their experiences in developing co-designed, technology-enriched, place-based SSI lesson plans throughout the PD program.

11.2 Implementation

11.2.1 Participants

The participants were 17 (16 female; 1 male) senior preservice teachers studying in the undergraduate biology teaching program at a university located in Istanbul, Türkiye. These students have completed their courses focusing on subject matter knowledge and biology teaching methods and voluntarily participated in the study's activities.

11.2.2 Professional Development Program for Preservice Teachers

In this study, an 11-week PD program was designed to teach SSI to preservice teachers. This PD program is intended to provide the necessary theoretical background of SSI and its aspects as well as to provide opportunities for preservice teachers to experience practical implementations of SSI in classroom practices through the co-design of lesson plans. The PD program consisted of two phases: phase one included the theoretical aspects, and phase two included the co-designing of the implementation and practices.

The first phase, which lasted for 6 weeks, included the theoretical part to introduce the foundational knowledge and critical concepts about SSI since the participants did not have any prior knowledge of this subject. The content included how SSI can be integrated into classroom practices, the importance and the guidelines of place-based SSI instruction, and a chance to observe and learn from real-time SSI classroom practices conducted by the researchers. These practices included an example of *vaccines* serving as a generic SSI implementation and a place-based practice on *water pollution* in Istanbul and its effects on the participants, since all the preservice teachers were the residents of Istanbul for the time being and, therefore, subject to exposure to the effects of this issue. The water pollution issue was selected specifically, as the participants live in the biggest metropole of Turkiye, the living space of almost ¼ of the country's population. Istanbul's drinking water quality is at risk of pollution due to urbanization, overpopulation, and industrialization (Coskun et al., 2008). Within this practice, under the guidance of the researchers, preservice teachers initially investigated their water consumption, their local water reservoirs' supply, the water quality, discharge limits, planned/unplanned water shortages as well as diseases that they and others suffered from water pollution, through gathering data mainly from Istanbul Water and Sewerage Authority (ISKI) and Ministry of Health. Then, they were asked to share their findings, daily habits, experiences, their reasoning about possible solutions, and how these issues can be integrated into learning environments in an open discussion format for a socioscientific discourse.

The second phase was the place-based co-design material development phase, in which preservice teachers were expected to design, develop, and revise place-based SSI teaching materials collaboratively, aiming for secondary school biology classes. In this phase, which lasted for 5 weeks, preservice teachers were divided into groups of 3–4 and worked collaboratively to co-design a place-based SSI lesson plan for 6–8 h on a unit from the biology curriculum integrating digital tools and technology and focusing on a local issue. During all these weeks in Phase 2, the researchers provided constructive and iterative feedback on the lesson plans. The content and the details of the PD program are demonstrated in Table 11.1.

11.2.3 Selected Topics and SSI Contexts: Biodiversity and Water Pollution

In this study, two SSI contexts, namely biodiversity and water pollution, were chosen by the preservice teachers to be integrated into the place-based lesson plans inspired by the local issues. The preservice teachers were free to choose the topics to design a lesson plan. In the following part, a brief introduction is presented about what these specific SSI contexts are about, their main components, and why they are important with regard to science education as well as locally.

Biodiversity was defined at the Convention on Biological Diversity (CBD) in 1992 as "the variability among living organisms from all sources including inter alia, terrestrial, marine and other aquatic ecosystems and the ecological complexes of which they are part, such as diversity within species, between species and of ecosystems" (Convention on Biological Diversity [CBD], 1992, p. 3) and since then, it's been considered as a complex system and interrelationships of the components of genetics, species, and ecosystems (Bermudez et al., 2021). Türkiye's geological location, with its bonds to both European and Asia continents, creates a great biodiversity from East to West with terrestrial, freshwater, and marine biodiversity (Şekercioğlu et al., 2011). This rich biodiversity allows almost over 9000 known native vascular plant species and many other endemic species to occur within the country. However, changing social and economic structures create great challenges and threats to this rich biodiversity in terms of conserving those crucial habitats.

Table 11.1 Content and the details of the professional development program

Phase	Week (hour)	Themes-content
Phase 1: Theoretical phase	Week 1 (2 h)	Explaining the purposes of the project Providing general information about the professional development program Introduction of digital web 2.0 tools such as Padlet, Canva, Google Tools, etc.
	Week 2 (2 h)	What are the socioscientific issues? The importance of SSI in science education and its relationship with national curricula What is SSI teaching? The multidimensional nature of SSI, theoretical frameworks of SSI teaching, the importance of place-based SSI instruction, guidelines for preparing a place-based SSI classroom practice
	Week 3 (2 h)	How can SSI be integrated into classroom practices by using argumentation, informal reasoning, and decision-making?
	Week 4 (1.5 h)	Providing a generic SSI classroom implementation on vaccines conducted by the researchers
	Week 5 (2 h)	Providing a place-based SSI classroom implementation on Istanbul's water pollution guided by the researchers
	Week 6 (1 h)	Reflections and discussions based on SSI implementations carried out on week 4 and week 5
Phase 2: Co-design phase	Week 7 (1 h)	Co-designing of the lesson plans: selecting the SSI topics focusing on a local issue from the biology curriculum and receiving feedback from the researchers
	Week 8 (1 h)	Co-designing of the lesson plans: Determining the sequence of activities for the unit and receiving feedback from the researchers
	Week 9 (1.5 h)	Co-designing of the lesson plans: Determining the content of the activities Deciding on which digital and Web 2.0 tools to be utilized and receiving feedback from the researchers
	Week 10–11 (1.5 h)	Finalizing lesson plans, receiving feedback from the researchers, and revisions

The Anatolia region is at risk of losing its biodiversity due to widespread threats such as uncontrolled urbanization, construction of dams, excessive irrigation, and wetland draining (Şekercioğlu et al., 2011). Moreover, it is known that using pesticides and fertilizers leads mainly to land and water pollution, eventually creating a risk to biodiversity and the whole ecosystem (Erisman et al., 2016).

On the other hand, another crucial element of the ecosystem and biodiversity, water is a critical natural resource and a habitat for many species and is also utilized for drinking and other developmental activities in human lives. Therefore, access to safe and clean water is imperative for human and other living organisms' health globally. However, due to many reasons, such as undesirable substances' occurrence in the water by discharging domestic and industrial effluent wastes, leakage from contaminated water tanks, marine dumping, radioactive waste, and atmospheric deposition, causing a decline in its quality and posing a threat to human and other organisms' health and eventually the environment. This contamination is called *water pollution*. It is known that water pollution, like other types of pollution (e.g., air, land), can create a great risk for each species' healthy lifespan, including humans (Moss, 2008; Haseena et al., 2017) as well as biodiversity. From the local perspective, according to the OECD, it is estimated that 20–50% of surface water in Türkiye is polluted by nitrogen, phosphorus, pesticides, and other hazardous substances (OECD, 2019). Similarly, Turkish seas are also constantly being polluted by household waste, land-based pollution, and the discharge of untreated waste (Turkish Marine Research Foundation [TMRF], 2018).

Several initiatives are being used to tackle these issues by using state-of-the-art technologies and methodologies. Organic farming is one of these initiatives that is a system that relies on natural methods and techniques to produce crops and livestock without the use of synthetic fertilizers, pesticides, genetically modified organisms (GMOs), or other harmful chemicals (US Department of Agriculture, 2023). The goal of organic farming is to preserve biodiversity, air, and water resources and to create a sustainable and healthy ecosystem that supports the well-being of both the environment and the people who depend on it.

As it can be inferred, with its complex nature with links to social, economic, political, and scientific aspects, biodiversity can be considered an SSI (Åkerblom & Lindahl, 2017). Similarly, according to Ewing and Sadler (2020), since both social systems (e.g., governmental and economic systems) and scientific systems (e.g., physical, life, and Earth systems) play a critical role in water pollution subject, it also creates a great space for studying it as an SSI. At that point, science education has a critical role in understanding the biodiversity and water pollution issues for raising the awareness of primary and secondary school students. From an SSI point of view, integrating biodiversity topics into the science and biology courses will promote students' understanding of their local and global environment and the intrinsic relationships between organisms and their ecosystems (de Freitas et al., 2023). In addition, adding local elements and issues, such as cases of biodiversity and types of pollution, will create authentic learning opportunities for students to be aware of their own environment, leading them to act proactively (Åkerblom & Lindahl, 2017).

11.2.4 Lesson Plans and Interviews

Lesson plans and interviews were the two main data sources of this study. The place-based lesson plans developed by the preservice teachers were the primary data sources. These lesson plans were co-designed and developed for 6–8 class hours through group work, with a requirement to incorporate technology and digital tools and focus on a local issue. The lesson plans encompassed a range of teaching materials, both online and printed, as well as in-class activities such as Web 2.0 tools, audio, videos, laboratory experiments, worksheets, presentations, handouts, classroom discussion questions, and evaluation rubrics.

The secondary data source of the study was the interviews. An interview protocol was developed using the existing literature and similar studies to investigate preservice teachers' experiences in the SSI material development process. After establishing the interview protocol's content validity, the individual interviews were conducted through online platforms with participants, each lasting about 40 min. Interview questions are listed below:

- How did technology integration contribute to your SSI lesson plans?
- What do you think about the importance of using technology in the classroom when addressing an SSI topic?
- Which technologies (Web 2.0) did you utilize in your lesson plans when addressing SSI?
- For what purposes did you use digital tools during the lesson process?
- Would it be possible to plan the SSI lessons without using technology? What would be the differences between those plans?

11.3 Outcomes

11.3.1 Co-designed Technology-Enriched Place-Based SSI Lesson Plans

Based on the characteristics of place-based education (Semken et al., 2017) that were introduced and the example implementation carried out to preservice teachers within Phase 1, the preservice teacher groups co-designed nine place-based lesson plans in total. The content of preservice teachers' technology-enriched, place-based, co-designed SSI lesson plans was analyzed regarding the utilization of these digital tools. This section presents the distinctive cases of the two preservice teacher groups' place-based SSI cases and the integration of the various digital tools into their lesson plans. As known, both biodiversity and water pollution issues have a local and global impact on the environment as well as on human life. Thus, two of the preservice teacher groups, Mustafa's group and Deniz's group (names are pseudonyms), each consisting of four preservice teachers, were selected for these specific SSI contexts, and they developed their lesson plans based on these contexts by also integrating digital tools. This section explains how these two local issues are embedded into the lesson plans via integrating digital tools as specific cases.

11.3.1.1 Case 1: The Case of Halil the Farmer on Organic Farming

For the biodiversity issue, preservice teacher Mustafa's group used a specific organic farming case from İzmir, a city in the Aegean part of the country. The lesson plan was prepared for 11th-grade secondary school biology class.

In their lesson plan, a local farmer named Halil's story created the main authentic space and content focus for a place-based SSI context. According to the story taken from a YouTube video, Halil was a farmer in İzmir, producing a variety of products in his yields and gardens, such as strawberries, mulberries, tomatoes, cucumbers, beans, pumpkin, etc., with traditional farming techniques by using certain types of pesticides, compounds, and fertilizers for extra yield within the shorter time period. One day, he realized that the pesticides he had been using were still effective at least 21 days after they were used, and those 21 days included the marketing and consumption of those products, which might cause serious health concerns for the consumers. Moreover, he realizes how the little insects were becoming resistant to those pesticides, which also threatened biodiversity as it harmed the environment. Then he decided to change his farming techniques; that was also the time when he was invited to a year-long certification program organized by Ege University, Faculty of Agriculture in İzmir, about organic farming along with the other 154 farmers in the region. However, during the course of the training program, 153 farmers dropped out due to difficulties with the techniques and extra effort required, except for farmer Halil. Halil completes the program with success, and then he starts to apply organic farming methods and techniques that he had learned in the program, such as using only organic substances from other types of ecologic waste and disposal instead of pesticides, preserving seeds with certain leaves for sustainability, etc.

Preservice teacher Mustafa's group used Halil's distinctive case as a starting point for an argumentative discussion, specifically his perseverance in the training program on organic farming in the name of preserving the ecosystem and the dilemma of the other farmers' preference to keep using traditional farming techniques even though they will harm the biodiversity and ecosystem. In a similar vein, it is known that organic farming is not merely an isolated agricultural practice; rather, it is a communal effort. Local farmers' markets, community-supported agriculture initiatives, and cooperative farming ventures create a network of shared responsibility and benefit, making the participants not just passive observers but integral contributors to the socio-scientific discourse surrounding organic farming in their locality. While İzmir, where Halil's story unfolds, may have a different agricultural landscape than Istanbul, the geographical proximity allows Istanbul residents to draw parallels and implications for their local farming practices. Also, for Istanbul residents, of whom the preservice teachers are part, the story of Halil holds significance as it sheds light on the potential health risks associated with farming techniques. Considering Istanbul's role as a major urban center, the awareness of sustainable and organic farming practices becomes crucial for the residents who are consumers of agricultural products sourced from various regions, including İzmir. All these features of the case had diverse meanings for the students as the consumers of food products, which Mustafa's group addressed in their place-based design. Thus, along with promoting a pro-environmental discourse and informal reasoning based on these issues, throughout the six consecutive lessons in their lesson plan, they also focused on how secondary school students' claims, warrants, backings, arguments, and counterarguments changed and evolved over data and interaction with peers. They also benefited from various digital tools to promote a deeper understanding of the issue within their design. These digital tools and purposes for inclusion are presented below:

- *YouTube*: YouTube was used as one of the digital teaching materials to introduce the main SSI context, the story of Halil, in a documentary video format.
- *Online T-Chart*: Mustafa's group used an online T-chart tool to provide a space for students to state their opinions and claims whether they agree or disagree, and why Halil's behavior is shifting to organic farming.
- *Nearpod*: Nearpod was used for various purposes in their lesson plan. The first purpose was to introduce the concepts of biodiversity, pollution, using pesticides, and other relevant concepts in a visually effective way. The second purpose was to organize the information the students gathered throughout their inquiry process and display it to the whole class at once.
- *Genial.ly*: The Genial.ly tool was used for online quizzes and real-time inquiry activity about organic farming on tomatoes.
- *Mentimeter*: A Mentimeter was used to get instant answers from students on the advantages and disadvantages of organic farming as pre and post-test.

11.3.1.2 Case 2: The Case of Water Pollution

Preservice teacher Deniz's group used a specific case of the factories' industrial effluent waste and discharge case and water pollution from their hometown cities, both located in the Southeastern part of the country. The preservice teachers stated that they specifically chose this topic to focus on due to their families' and relatives' exposure to the effects of the same issue in their home cities, based on the practices carried out within Phase 1 about Istanbul's water pollution. The lesson plan was prepared for tenth-grade secondary school Biology class.

In their lesson plan, scenarios related to water pollution created the main authentic space and content focus for a place-based SSI context. Firstly, a hypothetical scenario, which can create a dilemma, a case of building a textile and weaving factory in a people's residential area, was presented to students with a digital storytelling tool Powtoon. Along with this tool, in the following part, they used Canva, another interactive online graphic design tool, to show various contradictory visuals that can lead students to think critically about in favor of building a factory or against it. From there, they encouraged their students to construct their claims, data, warrant, backings, arguments, and counterarguments in an argumentative way based on this hypothetical scenario. They then asked their students to do an inquiry about the ecological balance and its related factors, mainly faulty utilization of water resources and water pollution, by using the internet and other relevant sources of information and using this information to support or rebut their arguments. These pro-environmental practices highlighted by Semken et al. (2017) aimed to enrich students' sense of place and to foster a better understanding of the specific environmental problem which they will study in the next step as the real cases happened in their region. The group then presented a real-life case from the Southeastern region of Türkiye, their home cities. In these real-life cases, two different factories built in the region started to discharge their industrial effluent waste into the region's clean water resources, resulting in eutrophication, which created a threat to both human and ecological life.

According to the news videos and visuals shared in the classroom, the water currents discharged with industrial effluent waste resulted in the water seeming in dark colors and people interviewed stated that there was a constant bad smell around the stream, which created serious health and ecological problems. Throughout the eight consecutive lessons in their lesson plan, they focused on how secondary school students' claims, warrants, backings, arguments, and counterarguments changed and evolved over data and interaction with peers by also using various digital tools on this specific topic. These digital tools and purposes for inclusion are presented below:

- *YouTube*: YouTube was used two-fold in their lesson plans. Firstly, it was to present the cases of the factories in those cities resulting in eutrophication and water pollution. Then it was also used to provide additional information about environmental pollution due to vast human consumption of products in an animated short film format.
- *Powtoon*: They used Powtoon, a digital storytelling tool, as the main material to present the hypothetical SSI case about building a factory.
- *Canva*: Deniz's group used Canva, an interactive online graphic design tool, to show various contradictory visuals that can lead students to think critically in favor of building a factory or against it.
- *Prezi*: Prezi was used to introduce the main concepts of air, land, water, noise pollution, radioactivity, eutrophication, their causes, and consequences in a visually effective way.
- *Mentimeter*: Mentimeter, a tool for interactive formative assessment activities that can also be used for creating instant word clouds, was used to create a word cloud on the concepts that they think are more important within the scope of the contradictory visuals presented.
- *Coggle Interactive Concept Mapping*: In the process of inquiry about ecological balance, Deniz's group asked their students to create an interactive concept map by using a Coggle tool based on the information that they gathered.

11.3.2 Digital and Web 2.0 Tools Used in the Lesson Plans

In this part, results related to the rest of the participants' place-based lesson plans regarding embedded digital tools are presented. Seven other lesson plans and the interviews with the preservice teachers showed that the preservice teachers benefitted and embedded numerous digital tools (19) with several purposes into their teaching. These purposes included creating an engaging learning environment, generating digital storytelling, creating interactive materials, accessing reliable information, drawing attention, gamification, and formative assessment. The main tools used and purposes related to that specific tool that they utilized are shown in the table below (see Table 11.2). It is noteworthy to indicate that all these listed tools are free-of-charge, user friendly, do not require an additional subscription fee for the demo users, and can be accessed through various devices such as mobile/smart phones, tablets, laptops, etc. via QR scan or key numbers. Within the interviews, preservice teachers widely referred to those crucial aspects of these digital tools and why they benefited from these tools within their lesson plans. They also stated within the interviews that technology-enriched teaching saves time and enables preservice teachers to focus on the student learning more. Moreover, the table shows that preservice teachers prioritized the "interaction and engagement" purpose

Table 11.2 Preservice teachers' use of digital tools and purposes

Digital tool/source	Purpose of use
Canva	Working interactively, cooperatively, and concurrently with other teammates
Classtools.net	Various purposes (name drawn, random group generation, online puzzles, games, wheels, Venn generators, etc.)
Coggle Interactive Concept Mapping	Formative assessment and interaction
EdPuzzle	Gamification and drawing attention
Genial.ly	Interactive presentation, animation creation, and formative assessment
Google Classroom	Working interactively, cooperatively, and concurrently with other teammates
Google Drive	Working interactively, cooperatively, and concurrently with other teammates
Kahoot	Interactive formative assessment
Mentimeter	Interactive formative assessment, word cloud generation
Nearpod	Working interactively, cooperatively, and concurrently with other teammates, gamification, and formative assessment
Online games	Drawing attention and formative assessment
Online T-chart	Interactive presentation, animation creation and formative assessment
Padlet	Working interactively, cooperatively, and concurrently with other teammates
Powtoon	Interactive presentation and animation creation, digital storytelling
Prezi	Interactive presentation and animation creation, digital storytelling
Reels videos	Informatic videos and audio for drawing attention
Search engines (Google, Google Scholar, Bing, etc.)	Accessing reliable information and relevant data collection
Smart Board	Engagement and interactive workspace
YouTube	Informatic videos and audio for drawing attention

among the other purposes for use. As known, these tools allow for multiple users to work simultaneously, which creates more opportunities for increased student engagement with the content.

11.4 Transfer to the Global Context

Both two SSI issues, *biodiversity* and *water pollution*, have local and global impacts and importance. In the twenty-first century, threats to biodiversity and different types of pollution (water, air, land, etc.) occur almost everywhere globally. Teachers around the globe are expected to raise awareness of these controversial issues as well as be proactive in creating authentic and effective learning opportunities to teach these issues at different levels. Place-based SSI instruction could be a beneficial tool to achieve these aims as it has a great potential to foster a better understanding of local and global environmental issues, raise awareness of different stakeholders, and promote students' sense of place. On the other hand, in this era, new digital tools for educational purposes are being released almost every single day and keep evolving. And especially as artificial intelligence (AI) technologies advance, teachers will be expected to integrate these new technologies into their teaching practices effectively. Thus, preservice teachers, as twenty-first-century citizens and future teachers, must stay updated with these new developments and changes. They should be able to effectively utilize these tools in their teaching practices, especially with SSI teaching. Teacher training programs are the main learning environments where future teachers could be supported with these initiatives to be prepared for these stages.

In line with this perspective, the technology-enriched co-designed place-based lesson plans created by preservice teachers during their teacher training years focused not only on the local problems but also included certain global aspects and impacts on the ecosystem benefiting certain digital tools. Moreover, incorporating these digital tools into place-based lesson plans could also allow secondary school students, who are the ultimate beneficiaries of these teaching materials, to engage more with the content. This integration would enable them to work collaboratively, thereby supporting their informal reasoning and decision-making processes. Additionally, it would encourage their pro-environmental and pro-societal practices through the utilization of various virtual tools. All these practices could promote a deeper understanding of the content as well as create an engaging and authentic learning environment. As a result, co-designed lesson plans' both local and global content as well as embedded digital tools' global accessibility enable all these materials to be applicable and adaptable for other countries' contexts. Moreover, the preservice professional development program model of this study focusing on improving both SSI understanding and improving SSI teaching practices of preservice teachers could also serve as an example to create engaging, interactive, collaborative opportunities for teacher training programs in other countries.

11.5 Conclusion

In this study, preservice teachers attended an 11-week PD program that intended to improve their SSI understanding and teaching practices by co-designing place-based technology-enriched lesson plans. It was found that preservice teachers embedded numerous digital tools into their lesson plans for various purposes. This study showed that as new tools were released and made available, preservice teachers' repertoire diversified in terms of including different tools with changing purposes (e.g., interactive formative assessment, word cloud generation, concept map creation, interactive presentation-animation creation, digital storytelling, gamification, using audio-video for drawing attention and working on an interactive-collaborative workspace) in which those even the current literature about technology integration to SSI, failed yet to state. It was also found that specific to SSI teaching, certain digital tools, namely Online T-Charts, Nearpod, Padlet, Powtoon, Coggle, and Genial.ly, were widely preferred and were used by almost every group. The main rationale for these tools to be included was that they provided an interactive and engaging space for students to state different opinions, share, refine, and organize ideas and thoughts easily, as well as helped them to monitor their changing informal reasoning during SSI implementations, and as a result of these, they were found to be suitable for inclusion.

In addition, from the co-design perspective, preservice teachers stated that at the beginning of the lesson plan design process, their expertise regarding the use of digital tools varied. Some had little or no experience at all, whereas others were quite experienced with them. However, at the end of the study, they stated that the co-design aspect of the study allowed them to learn from each other and increased their level of expertise in terms of incorporating digital tools into their lesson plans effectively. Moreover, it was also observed that co-design processes of lesson plans helped preservice teachers to understand the nature of SSI better since the design and development process themselves required engaging in iterative discussions, critical thinking, revisions, persuasion processes based on their ideas and arguments, and eventually reaching a consensus on the constructing of the learning activities. In sum, this study created a rich learning environment where the preservice teachers could build collaborative relationships with their peers, improve the quality of their tangible work with constant feedback, increase their level of expertise with digital tools, and integrate them into teaching practices. Yet, within the scope of teacher training programs, more studies are needed, especially about which works best in terms of various parameters of co-designed place-based practices on SSI understandings of pre-service teachers and the ways of effective integration of various state-of-art technologies into teaching practices.

References

Åkerblom, D., & Lindahl, M. (2017). Authenticity and the relevance of discourse and figured worlds in secondary students' discussions of socioscientific issues. *Teaching and Teacher Education, 65*, 205–214.

Alkhawaldeh, N. I., & Menchaca, M. (2014). Barriers to utilizing ICT in education in Jordan. *International Journal on e-Learning, 13*(2), 127–155.

Bermudez, G. M., Ottogalli, M. E., & García, L. P. (2021). Socioscientific argumentation: Toulmin's argument pattern for teaching and learning of biodiversity and its conservation. *Discourse and Argumentation in Science Education, 832–841.*

Borgerding, L. A., & Dagistan, M. (2018). Preservice science teachers' concerns and approaches for teaching socioscientific and controversial issues. *Journal of Science Teacher Education, 29*(4), 283–306.

Carson, K., & Dawson, V. (2016). A teacher professional development model for teaching socio-scientific issues. *Teaching Science, 62*(1), 28–35.

Cebesoy, Ü. B., & Chang Rundgren, S. N. (2023). Embracing socioscientific issues-based teaching and decision-making in teacher professional development. *Educational Review, 75*(3), 507–534.

Cebesoy, Ü. B., & Öztekin, C. (2016). Relationships among Turkish pre-service science teachers' genetics literacy levels and their attitudes towards issues in genetics literacy. *Journal of Baltic Science Education, 15*(2), 159.

Chang, H.-Y., Wu, H.-K., & Hsu, Y.-S. (2013). Integrating a mobile augmented reality activity to contextualize student learning of a socioscientific issue. *British Journal of Educational Technology, 44*(3), 95–99.

Christensen, R. (2002). Effects of technology integration education on the attitudes of teachers and students. *Journal of Research on Technology in Education, 34*(4), 411–433.

Convention on Biological Diversity (CBD). (1992). Retrieved from https://www.cbd.int/doc/legal/cbd-en.pdf

Coskun, H. G., Tanik, A., Alganci, U., & Cigizoglu, H. K. (2008). Determination of environmental quality of a drinking water reservoir by remote sensing, GIS and regression analysis. *Water, Air, and Soil Pollution, 194*, 275–285.

de Freitas, A. C., Do Nascimento, L. A., de Castro, R. G., Motokane, M. T., & Reis, P. (2023). Biodiversity and citizenship in an argumentative socioscientific process. *Sustainability, 15*(4), 2987.

Erisman, J. W., Eekeren, N. V., Wit, J. D., Koopmans, C., Cuijpers, W., Oerlemans, N., & Koks, B. J. (2016). Agriculture and biodiversity: A better balance benefits both. *AIMS Agriculture and Food, 1*(2), 157–174.

Ewing, M., & Sadler, T. D. (2020). Socio-scientific issues instruction. *The Science Teacher, 88*(2), 18–21.

Friedrichsen, P. J., Ke, L., Sadler, T. D., & Zangori, L. (2020). Enacting co-designed socio-scientific issues-based curriculum units: A case of secondary science teacher learning. *Journal of Science Teacher Education, 32*(1), 1–22.

Genel, A., & Topçu, M. S. (2016). Turkish preservice science teachers' socioscientific issues-based teaching practices in middle school science classrooms. *Research in Science & Technological Education, 34*(1), 105–123.

Han-Tosunoğlu, Ç., & İrez, S. (2017). Biyoloji öğretmenlerinin sosyobilimsel konularla ile ilgili anlayışları [Biology teachers' understanding of socioscientific issues]. *Journal of Uludag University Faculty of Education, 30*(2), 833–860.

Han-Tosunoglu, C., & Özer, F. (2022). Exploring pre-service biology teachers' informal reasoning and decision-making about COVID-19. *Science & Education, 31*(2), 325–355.

Hancock, T. S., Friedrichsen, P. J., Kinslow, A. T., & Sadler, T. D. (2019). Selecting socio-scientific issues for teaching: A grounded theory study of how science teachers collaboratively design SSI-based curricula. *Science & Education, 28*, 639–667.

Haseena, M., Malik, M. F., Javed, A., Arshad, S., Asif, N., Zulfiqar, S., & Hanif, J. (2017). Water pollution and human health. *Environmental Risk Assessment and Remediation, 1*(3). https://doi.org/10.4066/2529-8046.100020

Holmes, K. (2009). Planning to teach with digital tools: Introducing the interactive whiteboard to pre-service secondary mathematics teachers. *Australian Journal of Educational Technology, 25*(3), 351–365.

Kutluca, A. Y. (2021). An investigation of elementary teachers' pedagogical content knowledge for socioscientific argumentation: The effect of a learning and teaching experience. *Science Education, 105*(4), 743–775.

Kwon, S. M., Wardrip, P. S., & Gomez, L. M. (2014). Co-design of interdisciplinary projects as a mechanism for school capacity growth. *Improving Schools, 17*(1), 54–71.

Mishra, P., & Koehler, M. (2006). Technological pedagogical content knowledge: A framework for teacher knowledge. *Teachers College Record, 108*(6), 1017–1054.

Moss, B. (2008). Water pollution by agriculture. *Philosophical Transactions of the Royal Society B: Biological Sciences, 363*(1491), 659–666.

OECD. (2019). *OECD environmental performance reviews: Turkey 2019*. OECD Publishing. https://doi.org/10.1787/9789264087873-2-en

Özer, F., Erdaş Kartal, E., Doğan, N., Çakmakçı, G., İrez, S., & Yalaki, Y. (2018). Overview of professional development programs for teachers: Model, process, obstacle, theory and practice. *Amasya Education Journal, 7*(1), 63–97.

Ozturk, N., & Yilmaz-Tuzun, O. (2017). Preservice science teachers' epistemological beliefs and informal reasoning regarding socioscientific issues. *Research in Science Education, 47*, 1275–1304.

Saunders, K. J., & Rennie, L. J. (2013). A pedagogical model for ethical inquiry into socioscientific issues in science. *Research in Science Education, 43*, 253–274.

Şekercioğlu, Ç. H., Anderson, S., Akçay, E., Bilgin, R., Can, Ö. E., Semiz, G., et al. (2011). Turkey's globally important biodiversity in crisis. *Biological Conservation, 144*(12), 2752–2769.

Semken, S., Ward, E. G., Moosavi, S., & Chinn, P. W. U. (2017). Place-based education in geoscience: Theory, research, practice, and assessment. *Journal of Geoscience Education, 65*(4), 542–562.

Slavin, R. E. (2014). Making cooperative learning powerful. *Educational Leadership, 72*(2), 22–26.

Tondeur, J., Van Braak, J., Sang, G., Voogt, J., Fisser, P., & Ottenbreit-Leftwich, A. (2012). Preparing pre-service teachers to integrate technology in education: A synthesis of qualitative evidence. *Computers & Education, 59*(1), 134–144.

Topçu, M. S., & Çiftçi, A. (2023). Co-design and implementation of community-based engineering enriched science units: Exploration of pre-service science teachers' professional development. *Teaching and Teacher Education, 127*, 104095.

Turkish Marine Research Foundation (TMRF). (2018). *Pollution in Turkish seas*. https://tudav.org/en/our-fields/pollution/pollution-in-turkish-seas/

US Department of Agriculture – National Institute of Food and Agriculture. (2023). *Importance of organicagriculture*. https://www.nifa.usda.gov/topics/organic-agriculture#:~:text=Organic%20farmers%20utilize%20practices%20that,conditions%20and%20satisfy%20local%20markets

Voogt, J., Laferriere, T., Breuleux, A., Itow, R. C., Hickey, D. T., & McKenney, S. (2015). Collaborative design as a form of professional development. *Instructional Science, 43*, 259–282.

Voogt, J. M., Pieters, J. M., & Handelzalts, A. (2016). Teacher collaboration in curriculum design teams: Effects, mechanisms, and conditions. *Educational Research and Evaluation, 22*(3), 121–140.

Webb, N. M. (2009). The teacher's role in promoting collaborative dialogue in the classroom. *British Journal of Educational Psychology, 79*(1), 1–28.

Yapıcıoğlu, A. E. (2016). Fen bilimleri öğretmen adaylarının sosyobilimsel durum temelli öğretim yaklaşımı uygulamalarına yönelik görüşleri ve çalışmalarına yansıtmaları [The Views and reflections of preservice science teachers regarding implementations of socioscientific issue based instructions]. *Hacettepe Journal of Educational Research, 2*(2), 132–151.

Yates, A., Starkey, L., Egerton, B., & Flueggen, F. (2021). High school students' experience of online learning during Covid-19: The influence of technology and pedagogy. *Technology, Pedagogy and Education, 30*(1), 59–73.

Zorfass, J., & Rivero, H. K. (2005). Collaboration is key: How a community of practice promotes technology integration. *Journal of Special Education Technology, 20*(3), 51–67.

Chapter 12
Preservice Teacher Education for Place-Based Socioscientific Issues

Elif Selcan Öztay

12.1 Place-Based Education and Pedagogical Content Knowledge (PCK)

Place-based education recognizes the importance of local knowledge and real-world learning experiences in teaching and developing active citizens (Favier et al., 2021; Smith, 2002; Sobel, 2005; Subramaniam, 2020). Sobel (2005) defined place-based education as real-life learning experiences, emphasizing its benefits as increased student achievement and connection to local places, school development, parent and community partnerships, and engaged and enthusiastic teachers. This finding was consistent with Sternberg's (2003) study, which found that students who perform real-world tasks as experts and learn to think analytically and creatively can apply what they've learned to real-world situations. The primary goal of place-based education is "to ground learning in local phenomena and students' lived experience" (Smith, 2002, p. 590).

Although researchers have defined place-based education in a variety of ways, Smith (2002, p. 593) identified the following common features:

- Curriculum based on phenomena around the teachers and students.
- Learning experiences allow students to become the creators of knowledge rather than the consumers of knowledge created by others.
- Graduate school education is organized based on students' questions and concerns, so students take an active role in their learning process.

E. S. Öztay (✉)
Department of Mathematics and Science Education, Faculty of Education, Van Yüzüncü Yil University, Van, Türkiye
e-mail: selcankutucu@yyu.edu.tr

B. Namdar, E. Karahan (eds.), *Socioscientific Issues Focused Teacher Education*, Contemporary Trends and Issues in Science Education 60,
https://doi.org/10.1007/978-3-031-55233-5_12

- Teachers are experienced guides, co-learners, and brokers of community resources and learning possibilities. Teachers help students to gain skills to be effective learners.
- The wall between the school and the community becomes more permeable and frequently crossed. Hence, students can participate actively in the community enrolled in activities.
- Students' work is assessed in the same way as the work of any person is assessed in terms of contribution to the community; hence, students can perceive themselves as valued members of society.

Place-based educational pedagogical approaches encourage students' sense of place, ethical and sociocultural reflection, socioscientific reasoning, civic engagement, and understanding of scientific claims (Herman, 2018; Zeidler et al., 2019). These benefits could be obtained by participating in effective place-based SSI instruction, which includes higher-order practices like decision-making and reasoning, as well as pedagogical support (Herman, 2018). For example, Capkinoglu et al. (2020) studied middle school students' argumentation skills and participated in five local SSI. At the end of the study, they emphasized the importance of enriched place-based SSI instruction with structured scaffolding that makes concrete the complex situation, stating that "experiences with SSIs in their authentic locations are not sufficient by themselves for generating quality arguments; drawing conclusions and making logical judgments about the cases/effects/people encountered in those environments is also required" (Capkinoglu et al., 2020, p. 849). As a result, the effectiveness of place-based SSI is heavily influenced by teachers' teaching knowledge and instruction quality. Carter et al. (2014) also defined transformative learning theory as enabling individuals to recognize the underlying presuppositions and systems that underpin their beliefs and behaviors, emphasizing that "SSI [Socioscientific Issues] or STS [Science-Technology-Society] have emphasized issues-based teaching for generating awareness and participation" (p. 543).

12.1.1 Pedagogical Content Knowledge

Shulman proposed the concept of PCK in 1986 as a subcategory of content knowledge, which was later revised in 1987 as a unique blend of content and pedagogical knowledge. Shulman (1987) defined professional content knowledge (PCK) as a unique combination of content and pedagogy that teachers possess (p. 8). In other words, CK and pedagogical knowledge in terms of instructional practices, educational objectives, learner knowledge, assessment techniques, and classroom management were combined into one construct, PCK (Magnusson et al., 1999; Sande, 2010). Following Shulman's preliminary PCK model, many researchers have studied PCK, resulting in a variety of PCK models (Grossman, 1990; Magnusson et al., 1999; Gess-Newsome, 2015; Carlson & Daehler, 2019; Tamir, 1988). Science educators are very interested in Magnusson et al.'s (1999) model, and Magnusson et al. (1999) described the PCK transformation process as "of several types of knowledge for teaching" (p. 95). The PCK model includes five components: orientation toward science teaching (STO), knowledge of students' understanding (KoL), instructional strategies (KoIS), science curriculum (KoC), and science assessment (KoAs). STO is accepted as the overarching component among the five, influencing and being influenced by the others. In other words, teachers' science teaching orientation influences how they plan, implement, and evaluate their instruction and student comprehension. Figure 12.1 shows Magnusson et al.'s (1999) PCK model.

In 2012, PCK researchers gathered for the first PCK Summit, which built on a new model of teacher professional knowledge and skills, including PCK, known as the Consensus Model (CM) (Gess-Newsome, 2015). At the second PCK Summit in 2016, the CM was used to derive the PCK Refined Consensus Model (RCM). The RCM model includes three types of PCK: collective PCK (cPCK), personal PCK (pPCK), and enacted PCK (ePCK). Carlson and Daehler (2019, p. 82) define pPCK as "the personalized professional knowledge held by an individual teacher in science," whereas cPCK refers to "the specialized professional knowledge held by multiple educators in a field." ePCK is the "subset of knowledge that a teacher

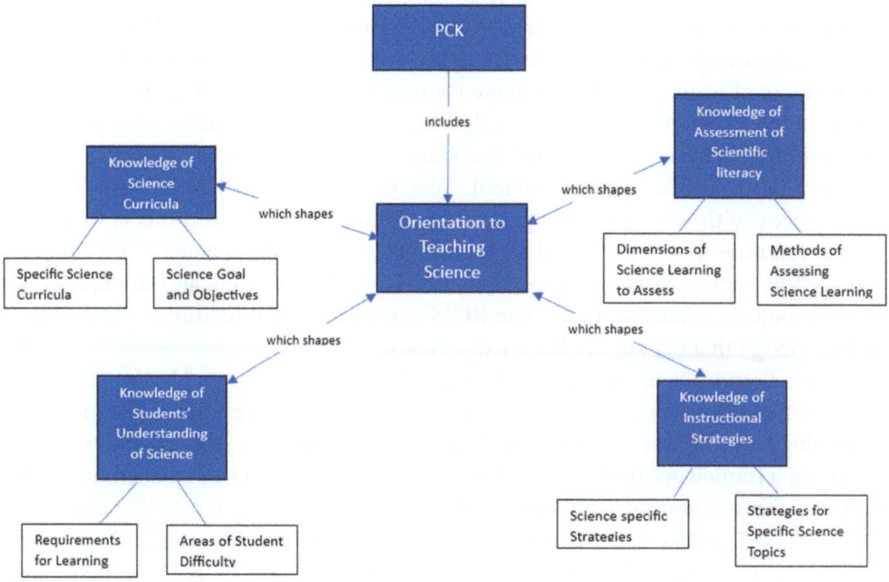

Fig. 12.1 Components of pedagogical content knowledge for science teaching (Magnusson et al., 1999, p. 99)

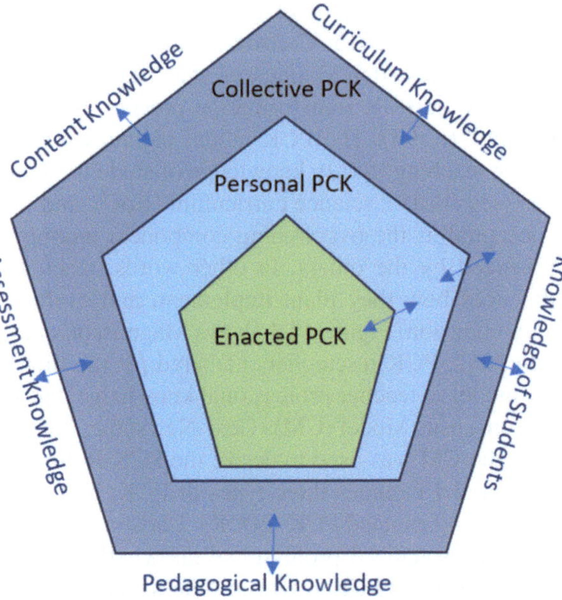

Fig. 12.2 Representation of the refined consensus model (RCM) of PCK (Carlson & Daehler, 2019)

draws on to engage in pedagogical reasoning while planning, teaching, and reflecting on a lesson" (p. 82). Figure 12.2 shows the Refined Consensus Model (RCM) of PCK. A two-way sign in the model denotes knowledge interaction between the three layers of professional knowledge for teachers: collective PCK, personal PCK, and enacted PCK. A flow of knowledge can occur when a teacher uses his or her enacted PCK to make instructional decisions or identify students' learning difficulties in relation to the specific content. Teachers' personal PCK is based on their experiences with teaching processes. Participating in research projects or professional learning communities that focus on contributing teachers' professional knowledge, such as student knowledge, assessment knowledge, and so on, can help to contribute to collective PCK. The RCM's most powerful feature is the exchange of knowledge that occurs via knowledge bases.

In the literature, researchers extended Shulman's pedagogical content knowledge framework to teachers who incorporate various science topics, disciplines, or technology into their instruction. For example, Mishra and Koehler (2006) proposed a framework that includes the essential qualifications of teacher knowledge required for integrating technology into their teaching, known as Technological Pedagogical Content Knowledge (TPCK). Aydin-Gunbatar et al. (2020) proposed a LESMeR model to help pre-service teachers develop their PCK for teaching integrated STEM education. As a result, different types of teaching knowledge are required to teach different topics in science; in this case, it is related to teaching SSI, which is known as PCK for SSI.

12.2 Pedagogical Content Knowledge and Socioscientific Issues

It is well known that incorporating SSI into instruction helps students understand the nature of science and motivates them to learn it (Bayram-Jacobs et al., 2019; Dori et al., 2003; Zeidler et al., 2005). Furthermore, using SSI effectively improves students' scientific literacy by connecting science to everyday life (Ke et al., 2021; Zeidler et al., 2005). Furthermore, SSI-based instruction helps students learn science content knowledge (Klosterman & Sadler, 2010; Sadler et al., 2016). An increasing body of research has shown that effective SSI instruction contributes to students' various types of knowledge and skills. For example, in the study conducted by Sadler et al. (2016), three teachers used SSI-based instruction to identify and treat sexually transmitted diseases using biotechnology. They discovered that SSI-based teaching can improve students' content learning. Aside from content knowledge, several studies show that teaching SSI helps students develop specific skills such as critical thinking, decision-making, moral reasoning, and argumentation (Sadler, 2004, 2009; Zohar & Nemet, 2002). Teachers should help their students use scientific knowledge and higher-order thinking skills to make informed decisions about SSI (Sadler, 2011).

Some researchers have developed a framework for teaching SSI (e.g., Eilks, 2010; Zeidler et al., 2011); however, a few researchers have focused on what teachers need to know to effectively embed SSI in their teaching (Bayram-Jacobs et al., 2019; Han-Tosunoglu & Lederman, 2021; Lee, 2016, 2018). In other words, few studies have looked at how teachers develop pedagogical content knowledge for SSI. There are few studies that look at how teachers' knowledge of SSI develops (Tidemand & Nielsen, 2017), particularly through the lens of PCK (Bayram-Jacobs et al., 2019; Han-Tosunoglu & Lederman, 2021). Several studies have investigated how in-service and pre-service teachers develop their PCK to teach SSI. The main questions remain to be addressed: "What is the teachers' model of SSI?" "What kind of knowledge should teachers have to teach SSI effectively?" "How is teachers' pedagogical content knowledge for SSI developed?". In response to these questions, researchers should create a framework for PCK for SSI and look into ways to improve teachers' PCK for SSI.

Transferring and embedding SSI into the science classroom has remained difficult, with related literature indicating that teachers struggled to use SSI during instruction (Han-Tosunoglu & Lederman, 2021; Pitiporntapin et al., 2016). Most pre-service and in-service teachers understand the importance of teaching in an SSI context, but they lack the necessary knowledge and pedagogy to do so (Anagun & Muhammet, 2010). Pre-service teachers' knowledge and understanding of SSI are limited and superficial (Espeja & Lagaron, 2015). Forbes and Davis (2008) also emphasized the importance of content and subject-specific knowledge in teaching-on-teaching SSI. Several research studies also found that training programs or teacher education method courses help pre-service teachers learn how to incorporate SSI into their instruction (Parr, 2013). For example, Espeja and Lagaron (2015) developed and implemented an SSI training program for elementary pre-service teachers to help them teach SSI. While pre-service teachers' initial conceptions of SSI were simple and limited, focusing on the topic and controversy, their final conceptions included more sophisticated ideas about uncertainty and argument dimension. Teachers have faced challenges in teaching SSI in their classrooms, including low teaching self-efficacy (Pitpiorntapin & Topcu, 2016; Tidemand & Nielsen, 2017), a lack of instructional time due to curriculum (Lee et al., 2006; Tidemand & Nielsen, 2017), a lack of instructional materials and tools (Ekborg et al., 2013; Lee et al., 2006; Sadler et al., 2017), and insufficient knowledge of science concepts (Kilinc et al., 2017). According to Lee et al. (2006), teachers understand the value of embedded SSI in science instruction. However, these factors prevented them from implementing SSI (Macalalag et al., 2020).

12.2.1 Pedagogical Content Knowledge for SSI

Because SSI consists of open-ended, ill-structured, and datable problems (Sadler, 2004), specific strategies and techniques for teaching SSI must be provided. SSI is defined as a framework that "focuses specifically on empowering students to consider how science-based issues and decisions made concerning them reflect, in part, the moral principles and qualities of virtue that encompass their own lives, as well as the physical and social world around them" (Zeidler et al., 2005, p. 360). SSI contexts in the classroom enable students to combine scientific knowledge with evidence-based thinking and reasoning. To assist and promote students, teachers must gain an understanding and pedagogical knowledge for implementing SSI instruction (Bayram-Jacobs et al., 2019). In other words, the higher the PCK for SSI, the more effective the instruction. As a result, teachers have a significant impact on how SSI is taught in school science (Forbes & Davis, 2008; Lee et al., 2006; Tidemand & Nielsen, 2017). Because PCK is topic- or domain-specific (Aydin et al., 2014; Mavhunga & Rollnick, 2013; Park et al., 2011), science teachers must have a certain level of knowledge about teaching and SSI in order to design a lesson plan and implement an SSI lesson (Bayram-Jacobs et al., 2019; Lee, 2022). Forbes and Davis (2008) emphasized that effective SSI teaching necessitates a variety of SSI-based teaching repertoires in addition to subject matter knowledge, or pedagogical content knowledge for SSI, because SSI and science topics differ. However, fewer studies have examined how teachers design and implement SSI in their classrooms, so research on teachers' PCK for SSI is still in its early stages (Bayram-Jacobs et al., 2019; Tidemand & Nielsen, 2017).

According to the literature, teachers had difficulty selecting appropriate SSI teaching materials, implementing alternative teaching strategies and techniques, and connecting to the science curriculum (Ekborg et al., 2013; Lee, 2022). Another challenge for science teachers during SSI instruction is the perception that the most important task is to teach scientific principles while viewing the social and ethical dimensions of science as a burden (Lee & Witz, 2009). Similarly, many teachers feel responsible for teaching science facts to students but believe that incorporating socioscientific teaching activities is optional (Forbes & Davis, 2008), or they are uncomfortable discussing ethical issues (Tidemand & Nielsen, 2017). In terms of assessment, teachers lacked knowledge about how to assess SSI learning (Tidemand & Nielsen, 2017). All of these issues can be resolved by improving teachers' PCK for SSI. More research is needed to develop in-service and pre-service teachers' understanding of PCK for SSI, including domains of teacher knowledge for teaching SSI and activities that contribute to developing PCK for SSI.

The theoretical framework of SSI-PCK was developed based on previous PCK literature (e.g., Magnusson et al., 1999; Shulman, 1986) and case studies on SSI teaching (Bayram-Jacobs et al., 2019; Han-Tosunoglu & Lederman, 2021; Lee, 2016, 2018). Bayram-Jacobs et al. (2019) defined SSI as a domain, and according to Shulman's (2015) view of PCK, SSI-PCK can be accepted as domain-specific teacher knowledge. They introduced four SSI-PCK components based on Magnusson et al.'s (1999) PCK model, which investigated how teachers' SSI-PCK developed and the relationship between those components. They had four components: knowledge of goals and objectives (GO), knowledge of students' understanding of science (SU), knowledge of instructional strategies (IS), and knowledge of how to assess students' understanding. Similarly, Lee (2016) proposed a model that closely resembled Magnusson et al.'s (1999) PCK model (Fig. 12.3). Figure 12.3 illustrates the model orientation for teaching. SSI is the overarching component that shapes the other five components. Orientation for teaching SSI is divided into two categories: teaching orientation and SSI orientation. Orientation types influence teachers' instruction. For example, SSI-oriented teachers used SSI to help students develop and apply critical thinking, decision-making, moral reasoning, and argumentation skills. On the other hand, teaching-oriented teachers used SSI as an effective tool for teaching science, engaging students' science interests and developing their knowledge of science content.

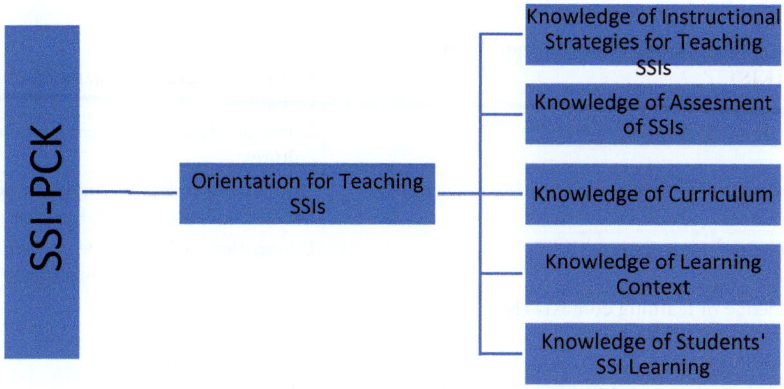

Fig. 12.3 Six SSI-PCK components and their relationship among these components (Lee, 2016, p. 542)

Han-Tosunoglu and Lederman (2021) proposed PCK-biological SSI (PCK-BSSI) and created an instrument to assess pre- and in-service science teachers' understanding of SSI, as well as their knowledge of how to teach SSI, using PCK. The items in the instruments were created using Shulman's PCK model (Shulman, 1986, 1987). The instrument is divided into three sections: teachers' background information, understanding of SSI, and PCK for BSSI. The third section focuses on teachers' understanding of teaching SSI, which includes curriculum, pedagogy, subject matter, student understanding, and school context domains similar to Shulman's PCK model.

In addition to Lee's (2016) SSI-PCK framework, Lee (2018) incorporated content knowledge from previous research. Teachers require a basic level of content knowledge, which includes science, technology, and the nature of science. The SSI-PCK framework is illustrated in Table 12.1.

Teachers' professionalism and instructional practices build on the knowledge gained during the teacher education program. Meaningful activities provided during this program help teachers believe and learn how to improve the quality of their teaching and student achievement (Chang & Park, 2020; Saeleset & Friedrichsen, 2021). PCK development is an ongoing process, and the first step is to offer a high-quality teacher education program. Teacher education programs prepare pre-service teachers for their future careers (Oztay & Boz, 2022). They gain foundational knowledge and skills that will help them develop their PCK during their teaching careers (Korthagen et al., 2006). Three major contexts have been for improving pre-service teachers' pedagogical content knowledge: teacher education courses, classroom teaching experiences, and certification programs. According to the literature

Table 12.1 SSI-PCK framework proposed by Lee (2018, p. 199, as cited in Lee, 2022)

Orientations for teaching SSIs (OTS)	Activity driven Knowledge and reasoning skill Application of science in everyday life Nature of science and technology Citizenship Activism
Knowledge of instructional strategies for teaching SSIs (KIS)	SSI lesson design Utilizing progressive instructional strategies Constructing collaborative classroom cultures
Knowledge of assessment of SSI learning (KAS)	Dimensions of SSI learning to assess Methods of assessing SSI learning
Knowledge of curriculum (KC)	Connection to the science curriculum Connections to other subject matter
Knowledge of learning contexts (KLC)	Classrooms Schools Community and society
Knowledge of students' SSI learning (KSL)	SSI learning experiences Difficulties in SSI learning SSI reasoning patterns

review, teaching education courses were the primary context that influenced pre-service teachers' knowledge of teaching SSI. For example, Sadler et al. (2017) proposed an SSI-based teaching and learning model. Macalalag et al. (2020) created a teacher education program that included a variety of activities based on case-based cultural practices, such as socioscientific discussions, hands-on activities, lesson plan development, and field studies. Similarly, Lee (2022) allows pre-service teachers to incorporate SSI into their teaching practices in order to improve their skills and knowledge of teaching SSIs.

12.3 Pedagogical Content Knowledge for Place-Based SSI

According to Smith (2002), teachers must chart their own path through the curriculum rather than simply implementing the mandated curriculum as a cookbook. This is entirely dependent on teachers' understanding of the curriculum. General curriculum models are ineffective for place-based education due to their unique characteristics. Smith (2002) proposed five approaches for teachers and educators interested in incorporating place-based education into their classrooms and schools: cultural studies, nature studies, real-world problem solving, internships and entrepreneurial opportunities, and introduction to community processes. The learning environment both inside and outside of school (for example, museums, schoolyards, and science centers) is critical to place-based learning (Smith, 2002; Subramaniam, 2020). Teachers should be aware of the importance of providing meaningful context for students to connect in-school and out-of-school science in order to learn science content. These learning environments provide opportunities for students to engage in real-world problems while applying their prior science content knowledge and connecting the science content to social awareness. Teachers should also be familiar with the main instructional strategies (e.g., inquiry, hands-on activities) and techniques that lead students to pose a problem (e.g., social or environmental problems), follow a procedure determined by teachers or students, and work independently or in groups to solve the problem. During instruction, the teacher serves as a guide, while the students are active learners who construct their own knowledge (Smith, 2002). Students gain ecological and social consciousness as they learn to conduct research, problem-solving, and decision-making strategies, as well as propose potential solutions to real-world problems (Subramaniam, 2020). Place-based education emphasizes the importance of exposing students to content knowledge and providing meaningful understanding through experiential learning (Smith, 2002; Sobel, 2005; Subramaniam, 2020). Meaningful understanding is achieved by connecting new knowledge to previous knowledge and experiences. Teachers should be aware of their students' prior learning experiences and learning difficulties related to content and place-based learning. In terms of assessment knowledge, teachers must focus on students' abilities, such as decision-making and problem-solving processes, as well as the quality of their decisions. SSI-based instruction requires students to propose arguments from various points of view, discuss their peers' arguments, and make a decision. Romine et al. (2017) proposed assessing students' understanding of SSI by presenting them with a new case scenario and observing their decision-making process.

Well-designed place-based teacher education can enrich teachers by creating lessons based on place-based education and teaching in ways that help students make sense of what they are learning by connecting with their local community (Chinn, 2012). During the teacher education program, pre-service teachers took place-based education courses, which included concrete examples of explicitly incorporating this into their instruction as well as learning instructional strategies to bridge in-school and out-of-school science. Developing teachers' knowledge for teaching is an ongoing process; therefore, teacher education programs for pre-service teachers and professional development programs for teachers should provide long-term training to capture changes in teachers' PCK for place-based education. Teachers with adequate place-based PCK serve as local experts or curriculum developers, emphasizing real places and issues related to local places during their instruction.

To summarize, the primary benefit of teaching SSI and place-based education is that it helps students develop relationships with their communities and regions. Teaching practice, well-designed professional development programs, and teacher education courses can help teachers teach place-based education and SSI. In other words, PCK for SSI and PCK for place-based learning. Further research could look into what practices could be incorporated into pre-service teacher education programs, as well as pre-service teachers' PCK development in place-based education and SSIs.

12.4 Pre-service Teacher Education for Place-Based SSI

As previously stated, place-based education benefits students. Place-based education connects students to their community and region (Smith, 2002), promoting active citizenship. At this point, teachers must internalize and integrate place-based SSI into their lessons. According to Capkinoglu et al. (2020), proper scaffolding is essential for effective place-based education. As a result, pre-service teachers should be prepared to deliver place-based SSI instruction during teacher education programs.

As previously stated, pre-service teachers have a superficial and limited understanding of teaching (Van Driel et al., 2002). However, Nilsson and Loughran (2012) stated that "That situation could change if preservice teachers were offered meaningful ways of defining, assessing, and explicitly developing PCK" (p. 700). As teacher educators, we should help pre-service teachers develop PCK during teacher education programs. Teacher education courses could help pre-service teachers improve their practical knowledge (PCK) of place-based SSI.

First, pre-service teachers received examples of effective placed-based SSI instruction in teacher education courses. This allowed them to experience the instructions from the perspectives of both a learner and a teacher. This would raise their awareness of the critical components of place-based SSI instruction. Herman et al. (2018) defined the critical components of place-based SSI instruction as

- Creating an authentic learning environment.
- Introduction of the issue at the beginning of the instruction.
- Providing experience and promoting higher-order skills (e.g., critical thinking, reasoning).

This is an excellent and detailed model for pre-service teacher education in placement-based SSI. Herman et al. (2020) proposed an SSI-infused, placement-based course for 24 university students representing various disciplines. The course consists of three parts: pre-departure instruction, field experience, and a post-field experience course. This course could be tailored for pre-service teacher education programs. In a recent study, Cebesoy (2023) developed 6 weeks of placed-based SSI instruction for pre-service science teachers focused on gold mining, inspired by Herman et al. (2020). Prior to the field experience, the researcher presented detailed information about gold mining, how gold is mined, and reports on the negative effects of gold mining. Pre-service teachers were also given an opportunity to hear citizens' perspectives on gold mining in the territory. During the field trip portion of the course, pre-service teachers interacted with stakeholders (e.g., geology engineers) and held group discussions in various parts of the mine. At the end of the field experience, pre-service teachers were able to evaluate different points of view and consider various perspectives on the local issue. During the post-field experience portion, the researcher presented reports and videos prepared by the Turkish Foundation for Combating Soil Erosion on the negative effects of mining. At the conclusion of the course, pre-service teachers were expected to reconsider their views on gold mining projects in their communities. They prepared a report outlining their reasoning for their evaluation. This type of instruction would be beneficial for pre-service teachers because it would help them learn how to implement place-based SSI instruction as future teachers while also improving their sense of place and attachment to their community. In another recent study, Avsar-Erumit et al. (2023) examined place-based SSI instruction on climate issues with pre-service teachers. They used field trips to give pre-service teachers experience in identifying climate issues. They used role-playing pedagogy to encourage pre-service teachers to consider other citizens' perspectives and apply their knowledge from local to global issues. This study was an example of a place-based pedagogical approach for pre-service teachers that aimed to improve students' understanding of the nature of science (NOS) and their civic values.

Second, after being introduced to place-based SSI, pre-service teachers were able to analyze lesson plans created by experienced teachers. As a result, they learned which points to emphasize when developing their lesson plan for instruction. Also, preparing lesson plans, either individually or in groups, could help them improve their teacher knowledge, or PCK on place-based SSIs. Content Representation (CoRe), developed by Loughran et al. (2004), could be used as a methodological tool to illustrate a teacher's or pre-service teacher's PCK on a specific science topic, particularly as a lesson plan. Pre-service teachers would benefit from preparing CoRe on place-based SSI during teacher education courses in order to link place-based SSI and pedagogy, which improves understanding of how to teach place-based SSI.

Aydin et al. (2013) and Van Driel et al. (2002) found that most teacher education studies emphasized classroom teaching to develop pre-service teachers' knowledge. After preparing CoRe or lesson plans for a place-based SSI, pre-service teachers were expected to teach this lesson in a classroom setting. In addition to classroom teaching experience, reflection and mentoring could be useful. After delivering a place-based SSI instruction, pre-service teachers could write a reflection on their classroom teaching experience in terms of both teaching and content. Reflecting on their perspectives enabled pre-service teachers to become more aware of their teaching experiences and redesign their instruction by taking into account the underlying reasons for their instructional decisions. Furthermore, mentoring could help pre-service teachers develop their place-based SSI PCK throughout the lesson plan/CoRe process and actual classroom teaching. The mentor could assist pre-service teachers in developing lesson plans, selecting appropriate pedagogy (e.g., role-playing, argumentation), and providing materials and alternative assessment methods for place-based SSI.

To summarize, providing examples of effective place-based SSI instruction through teacher education courses, developing structured lesson plans, such as CoRe, real classroom teaching, reflection, and mentoring could all be used to improve pre-service teachers' PCK development on place-based SSI through teacher education programs. There is a gap in the literature on studies that focus on developing pre-service teachers' knowledge of place-based SSI. In future research, new practicum courses focused on SSI and place-based SSI could be designed by combining strategies suggested by teacher education literature (e.g., preparing CoRe, reflecting on teaching experience, mentoring) to develop pre-service teachers' PCK on place-based SSI.

References

Anagun, S. S., & Muhammet, D. (2010). Teacher candidates' perceptions regarding socioscientific issues in science and technology instruction. *Social and Behavioral Science, 9*, 981–985.

Avsar-Erumit, B., Namdar, B., & Oğuz-Namdar, A. (2023). Promoting preservice teachers' global citizenship and contextualised NOS views through role-play activities integrated into place-based SSI instruction on climate issues. *International Journal of Science Education*. https://doi.org/10.1080/09500693.2023.2251189

Aydin, S., Demirdogen, B., Tarkin, A., Kutucu, S., Ekiz, B., Akin, F. N., et al. (2013). Providing a set of research-based practices to support preservice teachers' long-term professional development as learners of science teaching. *Science Education, 97*(6), 903–935.

Aydin, S., Friedrichsen, P. M., Boz, Y., & Hanuscin, D. L. (2014). Examination of the topic-specific nature of pedagogical content knowledge in teaching electrochemical cells and nuclear reactions. *Chemistry Education Research and Practice, 15*(4), 658–674. https://doi.org/10.1039/C4RP00105B

Aydin-Gunbatar, S., Ekiz-Kiran, B., & Oztay, E. S. (2020). Pre-service chemistry teachers' pedagogical content knowledge for integrated STEM development with LESMeR model. *Chemistry Education Research and Practice, 21*(4), 1063–1082.
Bayram-Jacobs, D., Henze, I., Evagorou, M., Shwartz, Y., Aschim, E. L., Alcaraz-Dominguez, S., et al. (2019). Science teachers' pedagogical content knowledge development during enactment of socioscientific curriculum materials. *Journal of Research in Science Teaching, 56*(9), 1207–1233.
Capkinoglu, E., Yilmaz, S., & Leblebicioglu, G. (2020). Quality of argumentation by seventh-graders in local socioscientifc issues. *Journal of Research in Science Teaching, 57*(6), 827–855.
Carlson, J., & Daehler, K. R. (2019). The refined consensus model of pedagogical content knowledge in science education. In A. Hume, R. Cooper, & A. Borowski (Eds.), *Repositioning pedagogical content knowledge in teachers' knowledge for teaching science* (pp. 77–94). Springer.
Carter, L., Rodriguez, C. C., & Jones, M. (2014). Transformative learning in science education: Investigating pedagogy for action. In L. Bencze & S. Alsop (Eds.), *Activist science and technology education* (pp. 531–545). Springer.
Cebesoy, U. B. (2023). Place-based SSI instruction: Current status and prospects. In O. Cardak & S. A. Kiray (Eds.), *Current studies in educational disciplines* (pp. 54–65). ISRES Publishing.
Chang, J., & Park, J. (2020). Developing teacher professionalism for teaching socio-scientific issues: What and how should teachers learn? *Cultural Studies of Science Education, 15*(2), 423–431.
Chinn, P. W. (2012). Developing teachers' place-based and culture-based pedagogical content knowledge and agency. In B. J. Fraser, K. Tobin, & C. J. McRobbie (Eds.), *Second international handbook of science education* (pp. 323–334). Springer.
Dori, Y. J., Tal, R., & Tsaushu, M. (2003). Teaching biotechnology through case studies-can we improve higher order thinking skills of nonscience majors? *Science Education, 87*, 767–793.
Eilks, I. (2010, September). *Making chemistry teaching relevant and promoting scientific literacy by focusing on authentic and controversial socio-scientific issues*. Presentation at the annual meeting of the society for didactics in chemistry and physics, Potsdam, Germany.
Ekborg, M., Ottander, C., Silfver, E., & Simon, S. (2013). Teachers' experience of working with socio-scientific issues: A large scale and in-depth study. *Research in Science Education, 43*(2), 599–617.
Espeja, A. G., & Lagaron, D. C. (2015). Socioscientific issues (SSI) in initial training of primary school teachers: Pre-service teachers' conceptualization of SSI and appreciation of the value of teaching SSI. *Procedia-Social and Behavioral Sciences, 196*, 80–88.
Favier, T., Van Gorp, B., Cyvin, J. B., & Cyvin, J. (2021). Learning to teach climate change: Students in teacher training and their progression in pedagogical content knowledge. *Journal of Geography in Higher Education, 45*(4), 594–620.
Forbes, C. T., & Davis, E. A. (2008). Exploring preservice elementary teachers' critique and adaptation of science curriculum materials in respect to socioscientific issues. *Science & Education, 17*(8), 829–854.
Gess-Newsome, J. (2015). A model of teacher professional knowledge and skill including PCK: Results of the thinking of the PCK summit. In A. Berry, P. Friedrichsen, & J. Loughran (Eds.), *Re-examining pedagogical content knowledge in science education* (pp. 28–41). Routledge.
Grossman, P. (1990). *The making of a teacher*. Teachers College Press.
Han-Tosunoglu, C., & Lederman, N. G. (2021). Developing an instrument to assess pedagogical content knowledge for biological socioscientific issues. *Teaching and Teacher Education, 97*, 103217. https://doi.org/10.1016/j.tate.2020.103217
Herman, B. C. (2018). Students' environmental NOS views, compassion, intent, and action: Impact of place-based socioscientific issues instruction. *Journal of Research in Science Teaching, 55*(4), 600–638. https://doi.org/10.1002/tea.21433
Herman B. C., Sadler T. D., Zeidler D. L., & Newton M. H. (2018). A socioscientific issues approach to environmental education. In G. Reis & J. Scott (Eds.), *International perspectives on the theory and practice of environmental education: A reader. Environmental discourses in science education* (Vol. 3, pp.145–161). Springer. https://doi.org/10.1007/978-3-319-67732-3_11

Herman, B. C., Zeidler, D. L., & Newton, M. (2020). Students' emotive reasoning through place-based environmental socioscientific issues. *Research in Science Education, 50*, 2081–2109. https://doi.org/10.1007/s11165-018-9764-1

Ke, L., Sadler, T. D., Zangori, L., et al. (2021). Developing and using multiple models to promote scientific literacy in the context of socio-scientific issues. *Science & Education, 30*, 589–607. https://doi.org/10.1007/s11191-021-00206-1

Kilinc, A., Demiral, U., & Kartal, T. (2017). Resistance to dialogic discourse in SSI teaching: The effects of an argumentation-based workshop, teaching practicum, and induction on a preservice science teacher. *Journal of Research in Science Teaching, 54*(6), 764–789.

Klosterman, M. L., & Sadler, T. D. (2010). Multi-level assessment of scientific content knowledge gains associated with socioscientific issues-based instruction. *International Journal of Science Education, 32*(8), 1017–1043.

Korthagen, F., Loughran, J. J., & Russell, T. (2006). Developing fundamental principles for teacher education programs and practices. *Teaching and Teacher Education, 22*(8), 1020–1041. https://doi.org/10.1016/j.tate.2006.04.022

Lee, H. (2016). Conceptualization of an SSI-PCK framework for teaching socioscientific issues. *Journal of the Korean Association for Science Education, 36*(4), 539–550.

Lee, H. (2018). *What is SSI education?* Parkyoung Story.

Lee, H. (2022). Pedagogical and epistemological challenges of pre-service science teachers teaching socioscientific issues: Based on the SSI-PCK framework. *Asia-Pacific Science Education, 8*(2), 301–330.

Lee, H., & Witz, K. (2009). Science teachers' inspiration for teaching socioscientific issues (SSI): Disconnection with reform efforts. *International Journal of Science Education, 31*(7), 931–960.

Lee, H., Abd-El-Khalick, F., & Choi, K. (2006). Korean science teachers' perceptions of the introduction of socioscientific issues into the science curriculum. *Canadian Journal of Science, Mathematics, and Technology Education, 6*(2), 97–117.

Loughran, J., Mulhall, P., & Berry, A. (2004). In search of pedagogical content knowledge in science: Developing ways of articulating and documenting professional practice. *Journal of Research in Science Teaching, 41*(4), 370–391.

Macalalag, A. Z., Johnson, J., & Lai, M. (2020). How do we do this: Learning how to teach socioscientific issues. *Cultural Studies of Science Education, 15*, 389–413.

Magnusson, S., Krajcik, J., & Borko, H. (1999). Nature, sources and development of pedagogical content knowledge for science teaching. In J. Gess-Newsome & N. G. Lederman (Eds.), *Examining pedagogical content knowledge: The construct and its implications for science education* (pp. 95–132). Kluwer.

Mavhunga, E., & Rollnick, M. (2013). Improving PCK of chemical equilibrium in pre-service teachers. *African Journal of Research in Mathematics, Science and Technology Education, 17*(1_2), 113–125.

Mishra, P., & Koehler, M. J. (2006). Technological pedagogical content knowledge: A framework for teacher knowledge. *Teachers College Record, 108*(6), 1017–1054.

Nilsson, P., & Loughran, J. (2012). Exploring the development of pre-service science elementary teachers' pedagogical content knowledge. *Journal of Science Teacher Education, 23*(7), 699–721.

Oztay, E. S., & Boz, Y. (2022). Interaction between pre-service chemistry teachers' pedagogical content knowledge and content knowledge in electrochemistry. *Journal of Pedagogical Research, 6*(1), 245–269.

Park, S., Jang, J. Y., Chen, Y. C., & Jung, J. (2011). Is pedagogical content knowledge (PCK) necessary for reformed science teaching?: Evidence from an empirical study. *Research in Science Education, 41*, 245–260.

Parr, J. C. (2013). *View of socioscientific issues among educators: The willingness of teachers to accept SSI into the classroom and the reason underlying those beliefs*. Unpublished doctoral dissertation. The University of Southern Mississippi.

Pitiporntapin, S., Yutakom, N., & Sadler, T. D. (2016). Thai pre-service science teachers' struggles in using socioscientific issues (SSIs) during practicum. *Asia-Pacific Forum on Science Learning & Teaching, 17*(2), 1–20.

Pitpiorntapin, S., & Topcu, M. S. (2016). Teaching based on socioscientific issues in science classrooms: A review study. *KKU International Journal of Humanities and Social Sciences, 6*(1), 119–136.

Romine, W. L., Sadler, T. D., & Kinslow, A. T. (2017). Assessment of scientific literacy: Development and validation of the Quantitative Assessment of Socio-Scientific Reasoning (QuASSR). *Journal of Research in Science Teaching, 54*(2), 274–295.

Sadler, T. D. (2004). Informal reasoning regarding socioscientific issues: A critical review of research. *Journal of Research in Science Teaching, 41*(5), 513–536. https://doi.org/10.1002/(ISSN)1098-2736

Sadler, T. D. (2009). Situated learning in science education: Socio-scientific issues as contexts for practice. *Studies in Science Education, 45*(1), 1–42. https://doi.org/10.1080/03057260802681839

Sadler, T. D. (2011). Socio-scientific issues-based education: What we know about science education in the context of SSI. In T. Sadler (Ed.), *Socio-scientific issues in the classroom* (pp. 355–369). Springer.

Sadler, T. D., Romine, W. L., & Topçu, M. S. (2016). Learning science content through socio-scientific issues-based instruction: A multi-level assessment study. *International Journal of Science Education, 38*(10), 1622–1635. https://doi.org/10.1080/09500693.2016.1204481

Sadler, T. D., Foulk, J. A., & Friedrichsen, P. J. (2017). Evolution of a model for socio-scientific issue teaching and learning. *International Journal of Education in Mathematics, Science and Technology, 5*(2), 75–87.

Sæleset, J., & Friedrichsen, P. (2021). Pre-service science teachers' pedagogical content knowledge integration of students' understanding in science and instructional strategies. *Eurasia Journal of Mathematics, Science Technology Education, 17*(5). https://doi.org/10.29333/ejmste/10859

Sande M. E. (2010). *Pedagogical content knowledge and the gas laws: A multiple case study.* Unpublished doctoral dissertation. University of Minnesota.

Shulman, L. S. (1986). Those who understand: Knowledge growth in teaching. *Educational Researcher, 15*(2), 4–14. https://doi.org/10.3102/0013189X015002004

Shulman, L. S. (1987). Knowledge and training: Foundations of the new reform. *Harvard Educational Review, 57*(1), 1–22. https://doi.org/10.17763/haer.57.1.j463w79r56455411

Shulman, L. (2015). PCK: Its genesis and exodus. In A. Berry, P. Friedrichsen, & J. Loughran (Eds.), *Re-examining pedagogical content knowledge in science education* (pp. 3–13). Routledge.

Smith, G. (2002). Place-based education, learning to be where we are. *Phi Delta Kappan*, 584–594. https://doi.org/10.1177/003172170208300806

Sobel, D. (2005). *Place-based education: Connecting classroom and community*. Orion Society.

Sternberg, R. J. (2003). What is an "expert student?". *Educational Researcher, 32*(8), 5–9.

Subramaniam, K. (2020). A place-based education analysis of prospective teachers' prior knowledge of science instruction in informal settings. *International Journal of Educational Research, 99*, 101497.

Tamir, P. (1988). Subject matter and related pedagogical knowledge in teacher education. *Teaching and Teacher Education, 4*(2), 99–110. https://doi.org/10.1016/0742-051X(88)90011-X

Tidemand, S., & Nielsen, J. A. (2017). The role of socioscientific issues in biology teaching: From the perspective of teachers. *International Journal of Science Education, 39*(1), 44–61.

Van Driel, J. H., De Jong, O., & Verloop, N. (2002). The development of preservice chemistry teachers' pedagogical content knowledge. *Science Education, 86*(4), 572–590.

Zeidler, D. L., Sadler, T. D., Simmons, M. L., & Howes, E. V. (2005). Beyond STS: A research-based framework for socioscientific issues education. *Science Education, 89*(3), 357–377. https://doi.org/10.1002/sce.20048

Zeidler, D. L., Applebaum, S. M., & Sadler, T. D. (2011). Enacting a socioscientific issues classroom: Transformative transformations. In T. D. Sadler (Ed.), *Socio-scientific issues in the classroom* (pp. 277–305). Springer.

Zeidler, D. L., Herman, B. C., & Sadler, T. D. (2019). New directions in socioscientifc issues research. *Disciplinary and Interdisciplinary Science Education Research, 1*(1), 1–9.

Zohar, A., & Nemet, F. (2002). Fostering students' knowledge and argumentation skills through dilemmas in human genetics. *Journal of Research in Science Teaching, 39*(1), 35–62. https://doi.org/10.1002/(ISSN)1098-2736

Chapter 13
New Directions for Place-Based Socioscientific Issue Instruction and Research

Bahadır Namdar and Engin Karahan

The objective of scientific literacy has been a focal point of science education policies for quite some time. Over the last 30 years, the field of science education has seen a strong push to increase scientific literacy due to the SSI trend. Students in progressive educational systems participate in democratic decision-making by having in-depth conversations about relevant, real-world issues, as stated by Zeidler et al. (2019). Therefore, SSI's curriculum and pedagogy are inherently contextualized because of the sociocultural framing that is inseparably tied to the quality of issues at the individual, community, and global levels. According to the pedagogical outcomes reported in this book's studies, more place-based, issue-focused content should be a priority for policymakers. This book discusses how teacher preparation programs can lay the groundwork for implementing place-based SSI pedagogies. Therefore, the policy changes can have a more substantial impact on a broader scale.

Standardized test scores have recently come to play a significant role in shaping educational policies on a national and international scale. Despite claims to the contrary in the SSI literature, high-stakes testing such as PISA or TIMSS presents unique challenges when assessing the efficacy of SSI curricula and instruction on a large scale (Zeidler et al., 2019). This is likely due to the fact that SSI is inherently contextualized. However, education for citizenship takes center stage in the new OECD report PISA 2025 Science Framework, which emphasizes the social context for assessing students' scientific competencies. However, many science education

B. Namdar (✉)
Department of Mathematics and Science Education, Faculty of Education,
Ege University, Izmir, Türkiye
e-mail: bahadir.namdar@ege.edu.tr

E. Karahan
Department of Mathematics and Science Education, Faculty of Education,
Middle East Technical University, Ankara, Türkiye
e-mail: enginka@metu.edu.tr

B. Namdar, E. Karahan (eds.), *Socioscientific Issues Focused Teacher Education*, Contemporary Trends and Issues in Science Education 60,
https://doi.org/10.1007/978-3-031-55233-5_13

programs around the world begin with the premise that the major purpose of science classes is to train the scientists of tomorrow. Nowhere in the OECD frameworks and related policy documents is the importance of teaching people to be scientists emphasized. The focus instead is on teaching the next generation to evaluate scientific findings critically and draw their own conclusions, skills that are expected to be useful throughout people's lives (Directorate General for Research and Innovation [European Commission], 2020). Therefore, while high-stakes testing does have an impact on science education policy, there is a clear trend toward science in the social context, which is closely related to the SSI framework.

Zeidler et al. (2019) did excellent work by addressing three broad and connected fields of study on SSI in the literature. First, the significance of socioscientific reasoning; second, the importance of taking a socioscientific perspective; and third, the role of informal and place-based contexts in SSI are all areas that could use more research. They expressed a strong interest in investigating the potential correlations between those trends and the utilization of SSI in various cultural settings. All those aspects (e.g., reasoning, argumentation, perspective-taking) were addressed at different levels in this collection of studies that focused on place-based pedagogies and their potential in teaching and learning SSI. Each chapter in this book addressed how to apply their findings to international contexts while maintaining a place-based perspective on SSI. As a result, comprehensive cross-cultural interpretations of place-based SSI pedagogies were offered. Claiming that the studies examining those trends drew from various sociocultural contexts is essential for comprehending the efficacy of SSI in different settings. We recognize that the tensions surrounding each SSI topic are socioculturally bounded. Our ability to present SSI topics alongside their related sociocultural contexts and then integrate them was, thus, of the utmost importance. Recent SSI studies, according to Zeidler (2015), have reflected international partnerships and centered on cross-cultural comparisons. This book offers a fresh perspective on the various ways SSI is understood across cultures through its overall structure.

The purpose of this book was to provide an overview of place-based pedagogies for teaching SSI. The book's chapters focused on specific outcomes of using place-based SSI interventions. These outcomes included but were not limited to, a sense of place, argumentation, reasoning, and scientific views on the nature of science. Each chapter's authors created an intervention based on place-based SSI strategies and tracked the changes in their participants. The book, in general, outlined prospective avenues and potential directions for pedagogy and research in place-based SSI.

The first key direction observed in the content of the book was the significance of place. Place-based education is an approach that directs students' attention to local culture, phenomena, and issues as the basis for at least some of the learning they encounter in school (Smith, 2017). In place-based education, "place" refers to the emotional and experiential connections between people and their built-environment social networks. Unlike space, which excludes people's thoughts, feelings, and memories, place incorporates them. Therefore, it is impossible to separate the human communities that inhabit a given location and the social and cultural elements they possess from any discussion of the concept of place. Self-identity is formed in relation to one's physical location (Tilley, 1994). According to the collected research in this book, students benefited greatly when exposed to place-based SSI that incorporated social and cultural perspectives. It was emphasized that a key factor in achieving successful outcomes was encouraging students to investigate the social and cultural dispositions of their immediate environment. Learners are more likely to engage in complex decision-making processes when they ensure socioscientific, personally meaningful, and relevant issues (Burek & Zeidler, 2015). Participants in the studies explored the sociocultural dynamics of their communities while investigating place-based SSI, allowing them to use higher-order thinking and reasoning skills. As a result, the more social and cultural aspects of the location were integrated into teaching and learning processes, the more complex skills learners acquired.

Another important message conveyed by the studies was the inevitability of fostering learner agency. Coulter (2020) argued that one of the most common limitations found in designs for place-based education is the need for better articulation of the ways in which we support meaningful agency among participants. He added that most place-based learning contexts are overly scripted by adults in charge, limiting opportunities for young people to connect with their local environment. As a result, it is reasonable to assert that agency is one of the understudied topics in place-based education. It is an even stronger case for place-based SSI. Only limited studies have addressed agency in the context of place-based SSI.

Taking advantage of the multifaceted and interdisciplinary nature of the SSI allows teachers to give students more agency in deciding how to approach the problem, as demonstrated in a study of a socioscientific agency by Karahan and Roehrig (2016). Although the studies in this collection did not specifically address agency as an outcome of their research, the interventions developed by the book's authors frequently included chances for preservice teachers to exercise agency by conducting their own research into and taking action on place-based SSI. Hence, the studies revealed the pedagogical affordances of agency in the settings of placed-based SSI interventions. It also opens up a new way to learn about agency in place-based SSI settings, along with all the difficulties and possibilities that come with them. Although the concept of agency is central to every school and after-school program, the term "agency" itself seems to have little traction outside of academic circles (Coulter, 2020). Hence, a meaningful way to introduce agency would be through place-based SSI.

To empower students to take appropriate, responsible, and effective action on issues of social, economic, environmental, and moral-ethical concern should be at the forefront of any science curriculum (Hodson, 2003). This is only possible if the links between school and society are strengthened. Empirical evidence also suggests that place-based education can facilitate partnerships between schools and communities and promote the use of local actors and resources (Place-based Education Evaluation Collaborative [PEEC], 2010). As a result, community engagement may be one of the potential directions for place-based SSI. Place-based pedagogies place a high value on community resources and emphasize ways to use them for rich and meaningful learning. Place-based education encourages local communities to actively identify and address SSI affecting their region, fostering a sense of belonging and empowering communities to work toward long-term solutions.

Strong partnerships between schools and other institutions, such as government agencies, non-profit organizations, and local businesses, can result from community involvement in place-based SSI education. These groups can work together to tackle difficult SSI by pooling knowledge and resources. This can be accomplished in a meaningful way through service-learning projects. Creating opportunities for students to put what they have learned in the classroom to use by participating in service projects that benefit their local communities. Students can learn from professionals in the field while also making a positive impact on their communities if they work with local organizations. Family participation is an integral part of community partnership. It is possible to increase parental and family involvement in schools by providing them with meaningful opportunities to do so. The placed-based SSI research should look into ways to strengthen this collaboration among various actors. It is critical to note that collaboration strategies must evolve as time passes. As a result, place-based SSI research and practice necessitate a dynamic process. Based on research, best practices in place-based SSI should center on interventions that fortify partnerships between educational institutions and their surrounding neighborhoods.

Each of the case studies in this book dealt with a regionally or geographically unique SSI. The authors made a great effort to make the problem scenarios relevant and interesting to the participants. According to their research, this enhanced the efficiency of the procedure. To give an example, rather than discussing nuclear energy as an SSI, Öztürk and Yenilmez Türkoğlu used a recently constructed nuclear power plant in Sinop to help their participants form more informed opinions and stronger arguments. Using local gold mining as an engaging issue, Cebesoy investigated how preservice science teachers' emotional reasoning evolved in response to place-based SSI instruction. Both examples, like the rest of the studies in the book, demonstrated the importance of situating a problem in its local context in order to achieve the goals of SSI. As a result, it is critical to emphasize that research should concentrate on developing more powerful frameworks to create locally strengthened SSI scenarios. Research on SSI should look into ways to incorporate place-based aspects into problem contexts.

One of the goals of science education has been regarded as encouraging young people's active participation in SSI. However, especially in societies with relatively low economic and welfare levels, the literature on youth participation and activism on SSI is slow to develop and extremely limited (Lewis-Charp et al., 2006). We cannot assume that the lack of research on youth activism will not have an effect on actual youth activism efforts. Children and young people need to be taught the knowledge and skills necessary to participate effectively in civic life (Levinson, 2010). In fact, there is great potential for youth activism to be bolstered by place-based SSI strategies. Research on SSI should explore strategies to promote youth activism. This collection of studies examined how place-based SSI could inspire learners to take an active role in solving local problems. Evidently, the place-based interventions in these studies encouraged their participants to investigate and resolve SSI actively.

As is the case with the vast majority of SSI research, the book's discussions of SSI centered on ecological concerns. Environmental stewardship, or the practice of acting and making decisions in a way that protects the environment for future generations, can result from such involvement. The book's emphasis on place-based education suggests that environmental stewardship can be given a more neighborhood-level focus. The term "local environmental stewardship" refers to "the actions taken by individuals, groups, or networks of actors, with varying motivations and levels of capacity to protect, care for, or responsibly use the environment in pursuit of environmental and/or social outcomes in diverse social-ecological contexts" (Bennett et al., 2018, p. 599). Science and environmental policies and programs worldwide have been paying more and more attention to and funding local environmental stewardship. More research is needed to determine how best to use place-based learning to encourage environmental stewardship in the context of SSI, given the importance of environmental stewardship in science education reforms and the results of the studies in the book.

One of the most cutting-edge trends in the literature on SSI is the emphasis on perspective-taking (Zeidler et al., 2019). It has been singled out as a must-have ability for a more scientifically literate populace. Despite the importance of the latter, Newton and Zeidler (2020) argued that it is difficult to operationalize the construct and identify effective pedagogical strategies to promote perspective-taking. When SSI is close to students' homes, they are more likely to approach them from multiple perspectives (Karahan & Roehrig, 2016). Hence, place-based SSI approaches are a good fit for improving learners' perspective-taking because they make issues more relevant to learners. While investigating place-based SSI, students can use perspective-taking skills as the perspectives they take are from the people they know. Students are better able to empathize and, therefore, take the perspectives of others when they are already familiar with those individuals or groups. However, further research should explore the potential intersections of place-based learning, SSI, and perspective-taking.

Finally, and perhaps most importantly, pre- and in-service teacher education is a significant area of research and practice in place-based SSI. This collection of research reports the results of place-based SSI interventions for preservice teachers. The study participants' reflections, however, suggested that they had never been through anything like this before. Despite the expanding body of research on place-based education and SSI, neither approach has yet been incorporated into traditional routes of teacher preparation. This is an even more significant case for place-based SSI education. Research should explore how to prepare pre- and in-service science teachers for using place-based SSI in their classrooms. Using place-based SSI approaches can be particularly challenging for educators working in centralized curriculum-based educational systems. Implementations rely heavily on their motivation, attitudes, and—most importantly—competencies.

References

Bennett, N. J., Whitty, T. S., Finkbeiner, E., Pittman, J., Bassett, H., Gelcich, S., & Allison, E. H. (2018). Environmental stewardship: A conceptual review and analytical framework. *Environmental Management, 61*, 597–614.

Burek, K., & Zeidler, D. L. (2015). Seeing the forest for the trees! Conservation and activism through socioscientific issues. In M. P. Mueller & D. J. Tippins (Eds.), *EcoJustice, citizen science and youth activism* (pp. 425–441). Springer.

Coulter, B. (2020). Developing youth agency through place-based education: Challenges and opportunities. In A. Cutter-Mackenzie-Knowles, K. Malone, & E. B. Hacking (Eds.), *Research handbook on childhood nature: Assemblages of childhood and nature research* (pp. 1501–1520). Springer.

European Commission, Directorate-General for Research and Innovation. (2020). *Science with and for society in Horizon 2020: Achievements and recommendations for Horizon Europe*. Publications Office. https://data.europa.eu/doi/10.2777/32018

Hodson, D. (2003). Time for action: Science education for an alternative future. *International Journal of Science Education, 25*(6), 645–670.

Karahan, E., & Roehrig, G. (2016). Use of socioscientific contexts for promoting student agency in environmental science classrooms. *Bartın University Journal of Faculty of Education, 5*(2), 425–442.

Levinson, R. (2010). Science education and democratic participation: An uneasy congruence? *Studies in Science Education, 46*(1), 69–119.

Lewis-Charp, H., Yu, H. C., & Soukamneuth, S. (2006). Civic activist approaches for engaging youth in social justice. In P. Noguera, J. Cammarota, & S. Ginwright (Eds.), *Beyond resistance! Youth activism and community change: New democratic possibilities for practice and policy for America's youth* (pp. 21–36). Routledge.

Newton, M. H., & Zeidler, D. L. (2020). Developing socioscientific perspective taking. *International Journal of Science Education, 42*(8), 1302–1319.

PEEC. (2010). *The benefits of place-based education: A report from the place-based education evaluation collaborative* (2nd ed.). Retrieved July 6, 2023, from http://www.litzsinger.org/PEEC2010_web.pdf

Smith, G. (2017, July 27). Place-based education. *Oxford research encyclopedia of education*. Retrieved July 11, 2023, from https://oxfordre.com/education/view/10.1093/acrefore/9780190264093.001.0001/acrefore-9780190264093-e-95

Tilley, C. Y. (1994). *A phenomenology of landscape: Places, paths, and monuments*. Berg Publishers.

Zeidler, D. (2015). Socioscientific issues. In R. Gunstone (Ed.), *Encyclopedia of science education*. Springer.

Zeidler, D. L., Herman, B. C., & Sadler, T. D. (2019). New directions in socioscientific issues research. *Disciplinary and Interdisciplinary Science Education Research, 1*(1), 1–9.